T0212547

Die blaue Stunde der Informatik

More information about this series at http://www.springer.com/series/15985

Walter Hehl

Chance in Physics, Computer Science and Philosophy

Chance as the Foundation of the World

 Springer

Walter Hehl
Thalwil, Switzerland

ISSN 2730-7425 ISSN 2730-7433 (electronic)
Die blaue Stunde der Informatik
ISBN 978-3-658-35114-4 ISBN 978-3-658-35112-0 (eBook)
https://doi.org/10.1007/978-3-658-35112-0

This book is a translation of the original German edition „Der Zufall in Physik, Informatik und Philosophie" by Hehl, Walter, published by Springer Fachmedien Wiesbaden GmbH in 2021. The translation was done with the help of artificial intelligence (machine translation by the service DeepL.com). A subsequent human revision was done primarily in terms of content, so that the book will read stylistically differently from a conventional translation. Springer Nature works continuously to further the development of tools for the production of books and on the related technologies to support the authors.

This Springer imprint is published by the registered company Springer Fachmedien Wiesbaden GmbH, part of Springer Nature.
The registered company address is: Abraham-Lincoln-Str. 46, 65189 Wiesbaden, Germany

Foreword

"All this strengthened my belief that one day I too would be able to be reconstructed from the material I would leave behind. It really can't be that difficult."
Clemens J. Selz, Austrian author,
Foreword by "Bot," 2018.

If reconstructing means pulling a "human-sounding artificial interview" from the material left behind, then that is not very hard. Uploading a human personality directly into a computer would be more like it, and is even impossible according to current understanding. But uniting a person's digital material—all emails, documents, publications, blog posts, and pictures—into digital identity and then using it for reconstruction is easily possible. Clemens Setz has demonstrated this with a "real" journalist asking questions, his digital diaries as a basis, and probably home-grown artificial intelligence. The artificial interviews—as if he were already in inaccessible past—became a massive sales success as the novel "Bot." However, "Bot" caused quite a furore in the professional literary scene.

The German-American computer scientist Joseph Weizenbaum (1923–2008) experienced a premonition of how easily people can be deceived by a computer with his program *Eliza* in 1966. His program was probably intended more as a satire on computers and the human relationship to them, but the simple bot[1] Eliza was seen—to the horror of the inventor—as a kind of clever and serious psychotherapist. Against expectations, *Eliza* was a great success and went down in the history of information technologies. Conversely, one might conclude the depth (or shallowness) of human conversations.

I had the pleasure of seeing Clemens Setz live on stage in an interview in September 2015. It became clear to me: if you write books about your fields of knowledge that are published, then you also deliver the basic material for (sometime later) an artificial *Walter Hehl*. At that time, I had already published three books:

---

[1] A *bot* is a computer program that performs tasks largely automatically without any human interaction.

- What I thought about current trends in information technology (2008)
- What I understood about the difficulties companies have when they want to innovate (2009, together with manager Rainer Willmanns).

These two books essentially grew out of professional work at IBM, and I actually owe them to the great company and fundamental pioneer of information technology, at least in the twentieth century.

Further in 2012, I tried to understand why we humans have such problems finding and understanding the truth, and how this process works:

- *Die unheimliche Beschleunigung des Wissens* is a story of the Copernicanizations of humanity: Astronomically, physically, and intellectually, we are being stripped of our specialness. We are becoming less and less unique (nor are we, as far as we know, in space).
- Then, as a missionary for the importance of software, in 2016 I wrote *Wechselwirkung*, a book that combines physics, philosophy, and software, which I see as a modernization of the worldview of the Austrian-British philosopher Karl Popper.
- After a detour to rectify the image of Galileo (2017) in public the foundations of religion from the perspective of physics, information technology, and psychology (2019).

From an emotional-human perspective, I have gathered personal memories that are important to me:

- In *Meine fünf Frauen* in 2020, I describe the most important women in my life. A small Sudeten German family history was created.

Behind all these books there is—I must confess—something evangelistic, something possibly unpleasantly instructive. The desire to point out something that is not known to everyone, but should be, is perhaps also a driving force for spending a year or two writing a book. But the books are actually, despite everything, the mainstream of knowledge and not exotic; only sometimes they contain (still) little-noticed truths. Take Galileo, for example, historians know that he tried to use tides to prove the heliocentric worldview, but that his proof was just nonsense. The laymen nevertheless consider him to be the one who "proved" the heliocentric system.

But by a stroke of personal luck, I still see a "missionary" gap in popular knowledge. The serendipity began with a celebratory dinner with the French-US mathematician Benoît Mandelbrot in 1986. He was my dining companion that evening. Mandelbrot was at that time my colleague in the company IBM. He had been a collaborator in IBM research since 1958 and was one of the most famous mathematicians of the twentieth century, though not of the usual abstract kind. He did pragmatic, multipurpose mathematics. Mandelbrot was the creator of the term "fractal" and the discoverer of the "apple man," the mathematical object named after him. All mathematically interested people know this little figure! It is perhaps the most complex structure we know. We will meet his mathematics in the chapter "Chance in nature" (without mathematical formulas).

As a developer in the IBM laboratory in Böblingen and as the head of the test laboratory of processor development, I calculated images of the Mandelbrot set as a hobby. I had a high-resolution printer at my disposal (an IBM 4250 printer which we had developed) and a park of computers (which were also developed in the Böblingen laboratory), with low computing power compared to today's computers, but with extended arithmetic accuracy, and I let them calculate for weeks. The accuracy could be used: It is a property of the Mandelbrot set that the complexity of the structures never stops; one can keep enlarging and delving deeper into the universe of numbers, and new structures keep appearing, often similar to things seen, but not identical. Some of the images (typically at $6000 \times 10,000$ pixels^2) went to an art exhibition, two images appeared in a book on the beauty of fractals by H.-O. Peitgen (Peitgen 1986), and I got to spend the evening with celebratory guest Benoît Mandelbrot.

I got to know his worldview. In his expression, nature is rough (and so is the financial world). By this, he means that smooth, regular, normal-geometric structures as learned in geometry at school are the exception. In addition, there is his idea of self-similarity, which I, as a physicist, can only follow to a limited extent. Self-similarity means that structures repeat identically or similarly at other scales. Physically, structures in other dimensions will only repeat themselves with systematically changed parameters (according to the scaling laws) as already Galilei showed. In mathematics, this possibly goes on indefinitely, smaller or larger. However, Mandelbrot set is only approximately self-similar; however, just this fact is exciting.

Indeed, "roughness" in its sense of "individuality of an object when looked at closely" is everywhere: in liquids, for example, turbulence; in botany, the different leaves on a tree; in humans, the structure of the iris; and in astronomy, the exact distribution of stars, for example, the shape of constellations. In nature, it is usually not the individual detail that matters, such as the ripples on a wave of water, but the laws that apply to all. Science has also devoted itself to this. Most of the time a point or an event does not matter much, but sometimes it does. Famous is the butterfly effect of the meteorologist Edward Lorenz in his work with the ingenious title:

"Can the flap of a butterfly's wings in Brazil cause a tornado in Texas?"

Incidentally, the effect is often misunderstood: The skier who accidentally triggers an avalanche (the "snowball effect") is not an example of the butterfly effect in the general sense; it is the clear catastrophic and unilateral borderline case. The danger was imminent beforehand, and the cause, as well as the effect, was clear. There was also no alternative for the avalanche; the direction of the avalanche is clear. We will discuss an extremely indecisive situation with Norton's Dome.

In evolution, a single case, a coincidence or mutation, could have caused something. This is even more true in human life; then chance even plays a decisive role. If we look closely, coincidence appears everywhere: even the precision of the celestial clockwork is lost when we look closely (and for a longer time). Chance appears indisputably and absolutely in quantum physics, but it is also visible to the naked eye. Yes, we even build machines to create randomness, such as the roulette wheel, or machines that use randomness internally, such as encryption machines.

In the foundations of world order and disorder, rule and chance, physics and computer science are mixed. Chance is related to causality, to the direction of time, to atomic structure, and to creativity and religion. To show this is the aim of the book. Chance brings into our otherwise clear and certain scientific world a great uncertainty of principle.

The following quote from Napoléon Bonaparte may not be an exaggeration at all:

> **"Le hasard est le seul roi légitime dans l'univers"**
> **"Chance is the legitimate ruler of the universe"**
> **Napoléon Bonaparte, French general, 1769-1821.**

The book intends to show that chance is built into the foundations of the world and is thus a third pillar of the construction of our world, alongside or in addition to the main pillars of physics and computer science. To do this, we trace an outline of the history of science, also as a history of chance, and of false and real certainty, and of just as much uncertainty. We try to show how much foreshadowing of modern concepts there was already in ancient science.

A comment on the quotes: I believe a field of knowledge includes the opinions and flashes of brilliance of others. Hopefully, it is a pleasure to read them and think along with them, maybe even refute them.

I learned an argument for citations from the director of IBM's research organization, Paul Horn, circa 2005:

> **"No matter how many good people you have, there are more outside."**
> **Paul Horn, US computer scientist, born 1944.**

It was meant as an argument to work with other research organizations. IBM research was then (and perhaps is now) the best industrial research organization in the computer field.

I think the same goes for quotes and the cultural world. No matter how many good thoughts you have, there are more outside.

Thanks to

I owe my first contact with the topic of chance in the history of science to a seminar on science "From Aristotle and Democritus to Newton" with Prof. August Nitschke in Stuttgart.

In short but intense meetings, I got impulses from my two IBM colleagues:

I learned the ubiquitous importance of chance from and with Benoît Mandelbrot. I learned about the concept of algorithmic complexity (and true randomness) in conversation with Gregory Chaitin. Mandelbrot and Chaitin are among the greatest mathematicians of the twentieth century.

The philosopher Klaus Mainzer has given me support with his work to see chance in its fundamental meaning. Mainzer is a tychist, although he does not use the word "tychism." What tychism means is explained in the book. The origin is the Greek goddess Tyche, the goddess of fate. For Mainzer, too, coincidence is itself the principle. And: Without chance nothing new comes into being.

I thank my wife Edith for her patience and thorough editing. Any linguistic errors in the text are due to my unchecked corrections and are entirely my responsibility.

Contents

Introduction: A Brief History of Science and Coincidence

"The beginning of all science is the astonishment that things are as they are."
Aristotle, Greek natural philosopher, 384 B.C.-322 B.C.

"Ignoramus et ignorabimus - We do not know and we will never know."
Emil du Bois-Reymond, German physiologist, speech in 1872.

"We must know, we will know."
Final sentence of a radio address and inscription on the mathematician's grave.
David Hilbert, German mathematician, speech in 1930.

"The answers you get depend on the questions you ask."
Thomas Kuhn, American philosopher of science, 1922–1996.

The title of this section is adapted from the wonderful book on the whole history of mankind by the Israeli philosopher and historian Yuval Noah Harari. The history of science is the hard part of the history of the world. "Hard" in the sense that the overall process of getting to know it is a random process with some leaps in the process, but the goal is quite clearly ("hard") given: The congruence of nature and mathematics. Also, the tool of universally repeatable experiments ensures correctness (most of the time, anyway). Nature enforces laws by its very nature, mathematics maps them sharply. The popular view *"everything is relative; you could have other science"* is nonsensical to the system of natural science. Actually, this already describes an important part of the book!

The history of the development of science is closely linked with the development of technology. This is quite natural, because knowledge creates power and technology in the form of weapons, means of production and products. This "knowledge is power" is itself probably the most famous saying in the history of science:

© Springer Fachmedien Wiesbaden GmbH, part of Springer Nature 2021
W. Hehl, *Chance in Physics, Computer Science and Philosophy*, Die blaue Stunde der Informatik, https://doi.org/10.1007/978-3-658-35112-0_1

"Scientia potentia est" "Knowledge is power".
Sir Francis Bacon, English philosopher, 1561–1621.

There is some confusion around the origin of the quotation in Bacon: Namely, the first occurrence has the form *"scientia potestas est"* and refers to God: Whose knowledge is his power. But we understand this in its later sense, Bacon 1620:

"Human knowledge and human power go hand in hand, for if the cause is not known, the effect cannot be produced."

The classic direction of formulation—from knowledge to power—has always applied in reverse in experimental science: From the ability to build the best experimental devices, the best science follows. One example—telescopes, from Galileo's two-inch telescope in the 1600s to today's 8 m or 10 m telescopes or the Hubble telescope in space.

But the reverse sentence also has a fundamental scientific meaning: From making and being able to make follows understanding. The baroque philosopher Giambattista Vico (1668–1744) introduced this constructive method of acquiring knowledge into philosophy with his principle:

"Verum et factum convertuntur - The true and the made are interchangeable."

Thus, "only that which we have made is knowable as true." This mantra of the philosopher Vico has fundamental meaning in the second pillar of our knowledge, computer science. A working program can prove, for example, that a material behaves as predicted by the finite element program[1] and its physical assumptions. The brilliant computer scientist Alan Turing introduced the method of the principle *"What we can do, we understand"* into computer science in 1950. It leads to the Turing Test named after him, the comparison of human ability with the ability of a computer: Can you have a quasi-human dialogue with a program? Even in Chinese?

There is a whole tower of similar tasks of increasing difficulty, all more or less solved:

Can a computer read script? More precisely: Can it read special, simplified print?
Can it read general print? Can it read handwriting it knows well? Can it read unfamiliar handwriting? Can he talk, for example, read something aloud? Can he write down a spoken conversation? A Chinese conversation? A Swiss-German conversation? Can he translate Chinese into English? Russian into German? Can he answer simple questions from a small field of knowledge? Can he answer general questions? Can he drive a car on the highway with little traffic? In heavy traffic? Can he have a natural general conversation? Can he diagnose a disease? etc.

---

[1] General digital method for calculating the properties of solids.

Nearly always the solvability of this task was doubted by many laymen (but also experts) at first, after solving then dismissed as a trivial matter until the next task. The author experienced this several times himself, e.g. *"a computer will never be able to drive a car"*—however 30 years ago. The above list is a small history of computers, but the development and the list go on, of course. All the projects behind these questions provided and continue to provide insights into the structure of human language, handwriting, how car driving works, how our brains work. Building a related successful program is proof of understanding a phenomenon.

In addition, there is another property and a fundamental difference. The classical analytical scientific method with observation and experiment (defined as tailored and limited observation) has led us into the depths of nature—the heavyweight limits of our horizon are Big Bang, dark matter, new elementary particles. The knowledge landscape up to this horizon is in principle well explored. The basic property of the method used is the investigation of causality. Knowledge is the understanding of causes; it is a bottom-up approach in the language of software.

These are the scientific sides of understandings, by causes or by construction. Psychologically (or polemically), two types or levels of understanding can be defined:

- The layman (and the classical philosopher):

    A process is understood when it can be grasped within the terms of normal life. This does not change if more noble expressions are used for it; it remains the normal world. Time runs evenly and space is not curved but Euclidean.

    But everyday terms and notions do not go far and need to be corrected again and again (Hehl 2016).

- The physicist says he (or she) has understood it (physically) when the process is understood in *corrected* terms of normal life. The correction may be, for example, that time stretches, space curves, or that the law of energy holds. It is the result of research. One simply gets used to accepting these corrected notions and thinking that way.

This definition is entirely in the spirit of the bon mot of the Hungarian-American mathematician John von Neumann (1903–1957), who said to his physicist friend Felix Smith:

"Young man, in mathematics you don't understand, you get used to it."

There is, of course, both the possibility that people want to express something impossible in colloquial language, as happens in some religions (for example, with the concept of the "Creator"), and, conversely, something simple in physicists' language. Thus, in esotericism and borderline areas of religion, terms from quantum physics are often used; this is suggested by the inherent mysticism of quantum physics. For instance, the Serbian-British physicist Vlatko Vedal (born 1971) says.

"[The vacuum of] quantum physics indeed aligns well with Buddhist emptiness."

The other scientific-technical method of understanding something is the reconstruction of this system property. With this constructive path, one starts conversely from the function as a whole. The direction of knowledge is in terms of software from top to bottom (bottom is the hardware), a top-down approach. Causality is replaced by teleology, the meaning of the whole.[2] By replicating functions, such as language, one gains an understanding of how that function is performed, what errors can occur, what other solutions are possible, and what improvements. Sometimes the understanding succeeds so well that, for example, a game is no longer a game at all, but only an algorithm. It is foreseeable that there will also be digital psychology and artificial souls to understand our feelings and mental defects.

Disturbingly, the building of systems seems to have no visible human or natural limits. But there is no natural law that sets hard limits here.

But let us go to the beginning. We want to look at essential phases in the history of science, information technology and chance together. For this purpose, we divide the history of science and information technology into three major periods: Antiquity, the Enlightenment, and modernity.

1.1   Ancient Science in today's Light

"May the study of Greek and Roman literature ever remain the basis of higher learning."
Johann Wolfgang von Goethe, German poet, published posthumously in 1833.

In science and philosophy, it is the Greeks and the study of their ancient science that teach us the beginning of it all.

As representatives of ancient science, we consider two philosophers and an astronomer and some of their teachings:

- Aristotle as the most important figure of ancient science until the scholasticism and the middle ages and reviled in the Enlightenment,
- Epicurus (or Democritus),
- Ptolemy and his practical scientific achievement.

Another philosopher, Plato, we will mention below as the origin of the "romantic" school of ideas.

Figure 1.1 shows the fresco "The School of Athens" by the painter Raphael da Urbino (1483–1520) in the Vatican. The painter himself is depicted (marked R on the far right) together with 21 of the most important representatives of ancient Greek philosophy and

---

[2]Teleology from the ancient Greek τέλος *télos* meaning 'meaning, purpose'.

Fig. 1.1 Peripatos—"The School of Athens" by Raphael (1510–1511). Fresco in the Vatican. The fresco glorifies ancient Greece as the cradle of culture. Aristotle is number 15, Epicurus is number 2, Ptolemy is number 20, and Plato is number 14. (Image: Wikimedia Commons, Bibi Saint-Pol)

science. The four persons mentioned, Aristotle, Epicurus, Ptolemy and Plato, are also included.

1.1.1 The Science of Aristotle

"If the state of the soul changes, this at the same time changes the appearance of the body, and vice versa: If the appearance of the body changes, this at the same time changes the state of the soul."
Aristotle, Greek natural philosopher, 384 B.C.-322 B.C.

From today's perspective, Aristotle's worldview is, on the surface, bizarre. For several centuries until the end of the Renaissance, however, it was a consistent, accepted system. Central to his physics is the theory of motion, which he derived largely from direct observation. Here are some statements along with "friendly" interpretations of Aristotle's laws from today's perspective:

- There are two areas of the sky with different laws, beyond the moon and below the moon.
  Compare the atmosphere on the one hand and interplanetary space on the other hand with vacuum. In the atmosphere a satellite burns up after some time, in the vacuum of space it remains almost indefinitely on its orbit. Aristotle, however, considers vacuum to be impossible.
- In the celestial sphere the planets run eternally on circles.
  See in the universe the celestial bodies move on conic sections, the circle is a special case.
- On earth there are natural and forced movements:
  "Naturally" a body tries to get to its natural place, the "fire" up, "heavy" down.

See the center of the earth as the center of gravity.

"Forced" movement comes about through a force on the body. Without force, the body remains stationary.

See the latter corresponds to a movement with friction on a rough surface.

- With a thrown stone or a projectile, he must introduce a curious auxiliary construction so that the projectile flies through the air: The air around the stone carries it on.

  This difficulty is striking and will be solved in the early middle ages with the impetus theory.

- A body falls faster the heavier it is.

  See this is correct for very light bodies and so called creeping motion, e.g. when a bullet falls into oil or at parachute jump.

- The motion of the planets is eternal—but it needs a beginning. Aristotle introduces a kind of abstract god who is invisible and does nothing else: the "unmoved mover".

  See the concept of this mover the launching of a satellite in space, which from now on runs freely and (almost) eternally on its orbit, if it was only brought high enough above the atmosphere.

The picture of Fig. 1.2 illustrates exactly the transition and thus the two world zones of Aristotle with the launching of satellites.

Fig. 1.2 Aristotle's two subworlds in one (Figure: Space and atmosphere. The image illustrates precisely the transition between the worlds with the launching of satellites. (Image: European Global Navigation Satellite System Agency (EGNSSA)/Pierre Carril)

Aristotle tried to integrate his observations and his everyday knowledge into a consistent system. The laws of nature are absolutely valid with him (so there are no miracles!). From his mechanics some of the ideas are transferable to modern times: For example, the part of the sky for one and the mechanics of a body when friction or viscosity dominate. Otherwise, his conception of mechanics earned him the derision of the Enlightenment.

To this is added the soul. For Aristotle it is the general life force, connected with the body, and thus mortal.

Compare with the current view (at least of the author) that life is a kind of running computer, that is, of "software" running on a material basis.

His idea of the holistic connection of body and soul is modern. It is more modern than Descartes'later division into here body, there mind. This dualism naturally fitted well with the hope of a fictional life separate from the body. With the universe eternally existing, the soul mortal, and the impossibility of miracles, it is surprising that Thomas Aquinas managed to integrate Aristotle, so rational, into the teachings of the Catholic Church in the thirteenth century—and for several centuries. But ecclesiastical, anti-Aristotelian notions of a Creator, of an afterlife e and of miracles have nevertheless become deeply ingrained in our collective understanding of the world.

Aristotle also discusses at length the role of chance. Thus he writes (Zekl 1986):

"The one (the chance event) has its cause outside of him, the other (the natural event) in itself."
Aristotle in "Physics, Lecture on Nature".

The natural event is completely determined by the laws of nature, the coincidence by "trivialities" or by "providence". The term "providence" means a subset of chance that exists only for humans, not for animals. But providence can be positive (a good fortune) or negative (a bad fortune). Thereby the providence is not put together by a higher being, but "it" adds, the providence is like every coincidence indeterminable in the causes.

For Aristotle, chance is not an explanation; only rules can explain. It can certainly not explain the world as a whole: How is the ordered world to arise from unordered chance? For him, the orderly movement of the planets is proof that there is order and perfection, at least beyond the orbit of the moon.

Considering Aristotle's starting point in the minimal and hazy knowledge of the fourth century BC on the one hand, and the wealth of consistent thought in him on the other, one has to say: chapeau. This is why the ridicule is unjust for the fallacy in women's teeth, popularized by the British philosopher Bertrand Russell. Here is his famous, somewhat disingenuous quote:

"Aristotle insisted that women had fewer teeth than men. Although he was married twice, it never occurred to him to recount it."

Aristotle relied on the knowledge of his time. He knew that stallions have more teeth, he knew that in the real world in humans the number of teeth varied, in women even more than in men. He did not disregard experience; he was on the receiving end of false or inaccurate information. Experience (as a prefiguration of experiment) was even at the heart of his natural philosophy. Here is a word from Charles Darwin, the British naturalist, in 1879:

"Aristotle was one of the greatest observers who ever lived."

1.1.2　The Ancient Atomists

"In reality, there are only the atoms and the void."
Democritus, Greek philosopher, 459 B.C.- 370 B.C.

The origins of atomistic Greek philosophies are the philosophical problems of divisibility of space, time and matter that arose one or two generations before Democritus.

- Divisibility of space and time:
  The best known is probably the turtle of Zeno of Elea or the paradox of Achilles and the turtle: Achilles cannot catch up with the turtle: In the time it takes him to catch up with the tortoise in each case, the tortoise in turn has got further, and so on. The problem will be solved cleanly only 2000 years later with the infinitesimal calculus. If there are physical limits for space and time, these are in any case many orders of magnitude smaller than today's measurability (Planck length and Planck time).
- Divisibility of matter:
  The atomists, such as the philosophers Leucipp, Democritus and Epicurus, do not consider matter to be arbitrarily divisible, for in dividing it one encounters indivisible particles, the atoms, which move in a vacuum.

It is an incredible premonition of reality: There are atoms! The premonition becomes tangible in chemistry in the nineteenth century, in the form of fixed relations between substances in chemical reactions (stoichiometry). Atoms become a scientific reality with Albert Einstein's work on Brownian motion and Ernest Rutherford's collision experiments.

Still at the beginning of the twentieth century the (excellent) Austrian physicist Ernst Mach sneered

"Have you seen any?" after the author and physicist Henning Genz.

as an answer to the question of the existence of atoms. Ancient atomic philosophy speculates further:

- The "ancient" atoms have the shapes of various regular geometric solids such as spheres, tetrahedra and cubes in different sizes.

See atoms are of different sizes, but their size is not sharply defined. Hydrogen is the smallest, Francium the largest. As single atoms they are spherically symmetric, in a compound they have different symmetries.

- The ancient atoms have hooks and eyes with which they connect to form bodies.

See the atoms can bond with other atoms, to form molecules or bodies. The number of links (bonds) is typical for the respective atom.

- The ancient atoms move in empty space.

See in gases, the atoms move in space and even consist largely of empty space themselves.

- The ancient atoms move chaotically.

See the motion of molecules in a gas or the Brownian motion of particles as in Fig. 1.3.

The Roman poet Lucretius describes the idea of chaotic motion very pictorially three centuries later: *"Atoms dance like particles of dust in a beam of light"*. It is chance in the modern sense. This is indeed how atoms and particles move, as we have known since the

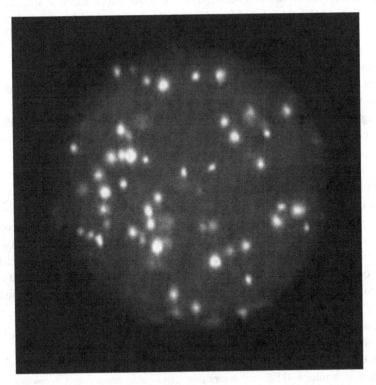

Fig. 1.3 Typical image of a Brownian motion, comparable to the motion of democritic atoms. (Image of the motion of particles of white ink in water under the microscope. (Picture: Faculty of Physics, LMU Munich)

nineteenth century. It is essentially the statement of the kinetic theory of gases of classical physics.

But in ancient atomism there is another forward-looking concept of chance. The clinamen. Clinamen is a spontaneous small change of motion in the flight of an atom. It must have been a curiosity to the earlier philosophers and non-physicists. Lucretius was ridiculed for it. Today this seems to be an ingenious trick to introduce purposefully "living" chance and to destroy boring simplicity. The German physicist Joachim Schlichting notes that this is the first time in western history that chance has been attributed a constructive role (Schlichting 1993).

Under the influence of gravity, the atoms would all fall parallel vertically without touching each other. That is why there are spontaneous fluctuations. Lucretius writes in *De rerum natura*—On the nature of things in the first century before Christ:

"... when bodies plunge straight down through the void with their own weight, then at fluctuating time and place from the course they leap off by a little, so that you are able to speak of changed direction."

A similar effect is well known in modern physics in a thought experiment, which goes back to the physicist and Nobel laureate Max Planck and one of the founders of quantum theory: It is about 1906 a carbon dust that upsets a too ideal order. We explain the idea in Chap. 5.

In other words, the clinamen and this Planck's carbon dust increase the entropy by leaps and bounds to the realistic equilibrium value.

There is no explanation for the clinamen, but if you are benevolent, you will find the reason for the trembling in quantum theory. It will explain this phenomenon with the "everything rushes" or "everything fluctuates". With randomness and the disordered movement of atoms, which can constantly recombine, it is not far to creativity. With this, there are thoughts in antiquity that are reminiscent of Darwin (though not in Aristotle, as Darwin assumed).

Ancient atomic theory is an approach reminiscent of the chemical phase of evolution. Molecules meet randomly to form larger groups: Complexity is built up.

Atomistic doctrine has yet another fascinating foreshadowing of the modern conception of the world: The existence of two pillars of the world, (inanimate) physics and (somehow animate) computer science. On the world model itself, below, more or at Hehl (2016). The two kinds of world show up in the doctrine of the two kinds of atoms, the robust atoms of the physical world and the subtle ones of the soul. The soul consists of a construct of particularly fine and light "soul atoms", similar to fire atoms. Thoughts are the movements of these soul atoms. When a person dies, the soul atoms dissipate and eventually join a new, emerging soul.

Admittedly it is daring, but is this not the foreshadowing of the two pillars "physics" and "information technology (IT)", of something physical, primary and something mental, secondary? "Mental" here in the sense of information-driven processes, or "software" for short (on hardware)?

A final remark on atomism. The philosophical idea that matter can be divided arbitrarily and indefinitely sees more the substance of a body as a whole. As a scientific idea, arbitrary divisibility has become nonsensical. Matter, if looked at sharply enough, becomes immaterial and dissolves into quantized fields. A pseudo-scientific doctrine still prevalent today assumes arbitrary divisibility: the homeopathy of Samuel Hahnemann (1755–1843). The "good" spirit of a substance remains effective at any dilution, the "bad" spirit is diluted away. The preferred dilution is 1 in 10^{60}. According to atomistic knowledge (not theory!) the dilution is so great that quite certainly not an atom or molecule is left in a globule of medicine! But for the founder of the doctrine in the eighteenth century the atomic doctrine was only deviant and forgotten philosophy.

1.1.3   Ancient Science Using Astronomy as an Example

"We consider it a good principle to explain phenomena by the simplest hypothesis."
Claudius Ptolemy, Greek astronomer and mathematician, 100-160 AD.

"Everything should be explained [made] as simple as possible, but not simpler."
Attributed to Albert Einstein, 1933.

The movement of the planets is not primarily a matter of uncertainty and coincidence but, on the contrary, of celestial order, albeit a rather complex order. The scientific task in ancient times is the calculation of the locations of the light points of the planets on the celestial sphere. It is mainly about the phenomenon, not about the cause! Of course, the earth is still in the center. The first major complexity that ancient astronomers had to solve was the apparent annual looping motion of the outer planets in the sky. To do this, they had to simulate the still unknown physics, namely the elliptical motion of the planets, i.e. the deviation from the circle with the existence of two foci each instead of a center and non-uniform orbital velocity.

Ptolemy had inherited from his predecessor Hipparchus an ingenious concept that could create loops: Epicycles, that is, he put circles on top of circles. Ptolemy moved some circles away from the center and with about 40 circles for the solar system (sun and the five planets) he got a model that would be used for 1500 years to predict the movements of the planets.

Figure 1.4 shows the observed positions of Venus (and Mercury) as seen from Earth over five years. First, the loops of the orbits show that epicycles fit the planetary problem in principle. To this end, the graph demonstrates a resonance phenomenon between Earth's orbit and Venus' orbit: Five Earth years correspond quite closely to eight Venus years. Otherwise, the course of the planets is a many-body problem in which everyone interferes with everyone else, and which looks quite chaotic if you look long enough. This resonance prevents the otherwise prevailing coincidence and thus keeps both orbital periods together until disturbances become too strong at some point.

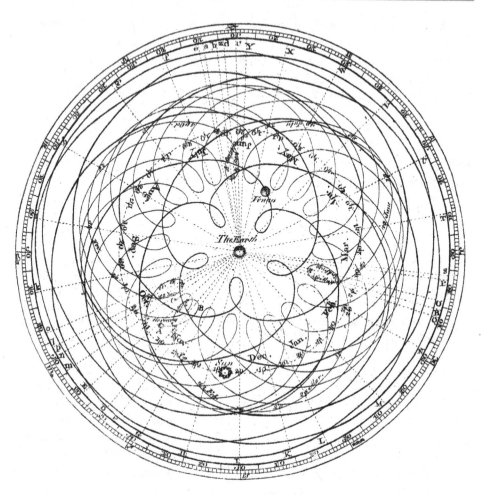

Fig. 1.4 The planetary orbits of Venus and Mercury as seen from Earth. (Image from the first edition of the Encyclopedia Britannica 1771 by James Ferguson after Giovanni Cassini. Note: This is not a diagram of epicycles. Image: Cassini apparent Wikimedia Commons, anonymous)

The method of Ptolemy is used for a millennium and a half, but it has a bad reputation in the sixteenth century. Errors crept in and its artificiality became increasingly visible. Copernicus reverses the basic position of the model and places the sun approximately in the center—but he uses the same procedure of epicycles, first 34 circles, later 40.

He has no proof that now in his construction the earth moves twice, around its own axis in about 24 h, around the sun in about 365 days. He expects better results as indirect proof. His results are depressing; they are worse than Ptolemy's.

Today it is clear why. Copernicus deliberately omitted an effective construct of Ptolemy, which he considered too artificial (the so-called equant).

Ptolemy and Copernicus both solve the "Platonic Axiom," the task Plato, set, they *"save the phenomena."* This expression meant in ancient astronomy to trace the complicated and mysterious orbits of the planets to a mathematics involving only circles and uniform circular motions, and thus to calculate them. These are numerical methods that the Church considers innocuous—until Galileo really means it with the sun at the center of the world.

Johannes Kepler will destroy the ancient premise of circles and be closer to physical reality. Mathematically the "destruction" is actually gentle, circles are after all a subset of ellipses. The astronomical models before are just geometric approximations and series evolutions in circles to reality, to ellipses. This makes Copernicus in particular the last ancient astronomer.

1.2   The Scientific Enlightenment

1.2.1   The Enlightenment in the Natural Sciences

"Nature and natural law were shrouded in night.
God said, 'Let there be Newton!' And the universe was filled with light."
Alexander Pope, English poet, 1688–1744.
Intended as an epitaph for Newton.

One of the precursors of the Enlightenment (and still with one foot in scholasticism) was the physicist Galileo Galilei (1564–1642). For Galileo there were two fundamental "books" of the world: The Bible and nature. Enlightenment was his assessment of this: The texts of the Bible could and should be interpreted in the spirit of the time in which they were written, but nature was unambiguous. The right to interpret the Bible was taken by the Church, and Galileo's conflict was thus pre-programmed. Incidentally, his main scientific argument in the dispute over whether the sun was stationary or moving was a completely false "proof" of the tides, which is now only a historical side note. Scientifically speaking, he should not have gotten involved in the trial (and the church, of course, should not have accused him).

Galileo was not the first to experiment and measure. But his experiment of letting balls roll down an inclined plane arbitrarily slowly and thus easily measurable instead of throwing them from a tower was indeed ingenious and the result unambiguous. In addition, like the Greek philosopher Plato, he emphasized that nature was written in the language of mathematics. However, for him mathematics was simple geometry and rule of three. He did not write a single equation. Overall, as an artist and experimenter, he is a late Renaissance man, similar to Leonardo da Vinci, and not an Enlightenment man. For more on this, see Hehl (2018).

The German astronomer Johannes Kepler (1571–1630) is close to the beginning of the Enlightenment with his discovery of the planetary laws and his mathematical work, but not yet as a person. The observations of the astronomer Tycho Brahe are so accurate that

Kepler cannot obtain the true orbit of Mars using the ancient method with circles on circles in manageable numbers. According to the US physicist and philosopher Thomas Kuhn (1922–1996), such a situation represents a paradigm shift, of which Kuhn says:

> **"Although the world doesn't change when the paradigm shifts, scientists work in a different world afterward."**

It is even possible to specify a meaningful day for the paradigm shift: May 15, 1618. According to Kepler, it is the day of the discovery of the third law named after him. Figure 1.5 shows the law graphically: The cubes of the orbital axes divided by the squares of the orbital periods give a single number. This had to seem like magic. The discovery has nothing to do with Plato and Ptolemy and antiquity. May 15, should be celebrated as World Science Day!

It is a mystery (to me) how Kepler was able to perform his extensive calculations of the Martian orbit without a computer, only with logarithms and in adverse living conditions! As a person Kepler was not an enlightener, rather a mystic. He believed in the harmony of the world, which was to be discovered. Perhaps he even believed in astrology, at least in real horoscopes, as he drew them up himself. From today's point of view, however, it is precisely his holistic view that is sympathetic and attractive.

Kepler already found his laws on the basis of astronomical observations with the naked eye, albeit thanks to the extraordinary observer Tycho Brahe. But it took two simple inventions from the Netherlands to bring about a tangible paradigm shift: The telescope

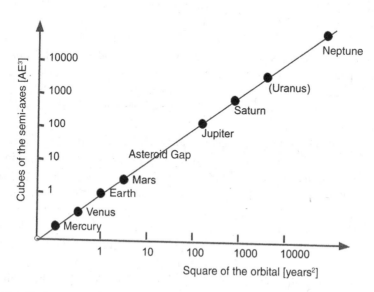

Fig. 1.5 Graphical representation of Kepler's third law. (Image: From "Galileo Galilei kontrovers", Walter Hehl 2017, Springer Vieweg)

and the microscope. With these instruments, man transcends the limits of his senses and sees hitherto unseen and undreamed-of things.

Up to these inventions the world of knowledge had been satisfactorily closed, secured by the words of the Bible and the writings of Aristotle. According to these, the smallest botanical object is the mustard seed and the smallest animal is the flea; no one would think of looking for anything smaller. Indeed, Galileo reports of contemporaries who refused to look through the telescope on this pious grounds!

But now, in a drop of water from the pond, you see a multitude of teeming little animals, and in the night sky there are a multitude of stars that can only be seen through the telescope: What are they for, if we humans cannot see them after all? Surely the world is only for us? Also, the moon has real mountains and is similar to the earth and no longer heavenly-perfect. How then does it remain in the sky?

There is a great uncertainty, which is resolved in the next centuries. It begins with Isaac Newton and his book "*Philosophiae Naturalis Mathematica Principia*", in which the basic concepts and fundamental laws of mechanics are described. Force, inertia, momentum and energy are defined and Kepler's laws are derived. These terms had been used in a confused manner for several hundred years. Newton finds the general law of gravitation: Everything attracts everything. He is thus able to derive Kepler's laws in a sharper form—the circle is closed. But Newton is only a rational physicist during the day, at night he does alchemy or simply chemistry, he even searches for the philosopher's stone. He is a mystic, darker than Kepler. In this sense the saying of the owner of Newton's letters about alchemy is valid:

"Newton was not the first enlightener, he was the last wizard."
John Maynard Keynes,
Economist and amateur historian, 1946.

But still, it is actually legitimate to want to change one element into another. Alchemy is a precursor of chemistry and is a single mystery. Enlightenment in chemistry no longer reaches Newton:

- The French chemist Lavoisier understood combustion as a reaction with oxygen in 1783 and thus banished the false phlogiston theory,
- the integer weight ratios in various chemical compounds observed Dalton 1808,
- the periodic system of elements will create the Russian Dimitri Mendeleev and the German Lothar Meyer in 1869 independently of each other.

Of course, we do not understand why and how atoms combine to form molecules, and why the Periodic Table of elements has this particular structure. But the understanding will be enough to make chemistry a science and to build up a flourishing chemical industry. By the end of the nineteenth century, people will feel they have understood everything important.

Classical physics was particularly successful in explaining the properties of gases: A gas, enclosed in a space, is tamed randomness. However, it is about the behavior of very,

very many randomly moving particles. The philosophical climax is the concept of entropy as the sweeping measure of disorder and order.

Even more successful is the physics of the nineteenth century in the field of electromagnetism, whose basic laws were found and describe, for example, induction and electromagnetic waves: Light is an electromagnetic wave. But there is a peculiarity here. The Scottish physicist James Maxwell established the basic equations of electromagnetism in 1865—and his equations are more than "enlightening". They are "relativistic" even before the discovery of relativity, i.e. they conform to relativity theory, and they are the basis of electrodynamics that is still valid today.

The greatest scientific achievement of biology here could be the discovery of evolution in the nineteenth century. But evolution is in essence not a topic of physics, but closer to information technology. It is a software technology of nature and thus belongs in principle to the second pillar of the world. Digital software only emerges with the great computer systems in the modern age.

1.2.2 The Dawn of Information Technology

"It is unworthy to waste the time of excellent people in menial arithmetic, because in using a machine even the most simple-minded can write down the results with confidence."
Gottfried Wilhelm Leibniz, German polymath, 1646-1716.

The history of the second world pillar is above all a history of action, technology and invention. The age of Enlightenment begins to germinate what is to become the key technology in the twenty-first century and forms the second pillar in the world model alongside physics, information technology. The beginning is mechanical and not very intellectual. The first calculating machines were designed and even built, for example, by the Swabian astronomer Wilhelm Schickard in 1623 and by the French philosopher Blaise Pascal in 1642.

The philosopher Arthur Schopenhauer (1788–1860) draws a false conclusion from the possibility of being able to perform basic arithmetic mechanically: He therefore, considers mathematics to be a menial activity without profundity. This is totally wrong: The triviality applies only to the individual operations—but from them, one can build arbitrarily complex systems that can read, speak and recognize faces!

Mechanically, it is difficult to build more complicated calculating machines with many parts working together like gears (the required accuracy in manufacturing is high), but milestones are already the designs of a "Difference Machine" (1822) and especially the "Analytical Machine" (1837) by the English mathematician Charles Babbage (1791–1871). It is the first real computer design with an arithmetic unit, control unit and internal memory, which, however, was never realized.

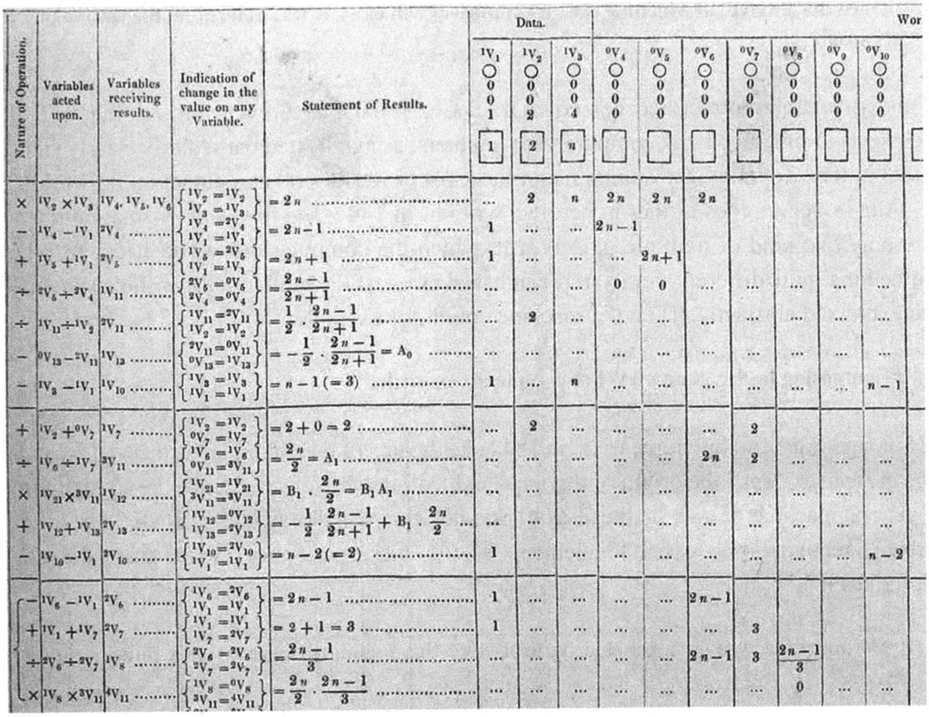

Nature of Operation.	Variables acted upon.	Variables receiving results.	Indication of change in the value on any Variable.	Statement of Results.	Data.			Wor						
					$^1V_1$ ○ 0 0 0 1 [1]	$^1V_2$ ○ 0 0 0 2 [2]	$^1V_3$ ○ 0 0 0 4 [n]	$^0V_4$ ○ 0 0 0 0 [ ]	$^0V_5$ ○ 0 0 0 0 [ ]	$^0V_6$ ○ 0 0 0 0 [ ]	$^0V_7$ ○ 0 0 0 0 [ ]	$^0V_8$ ○ 0 0 0 0 [ ]	$^0V_9$ ○ 0 0 0 0 [ ]	$^0V_{10}$ ○ 0 0 0 0 [ ]
×	$^1V_2 \times {^1V_3}$	$^1V_4, {^1V_5}, {^1V_6}$	$\left\{ {^1V_2} = {^1V_2} \atop {^1V_3} = {^1V_3} \right\}$	$= 2n$	...	2	n	$2n$	$2n$	$2n$				
−	$^1V_4 - {^1V_1}$	$^2V_4$	$\left\{ {^1V_4} = {^2V_4} \atop {^1V_1} = {^1V_1} \right\}$	$= 2n - 1$	1	...	...	$2n - 1$						
+	$^1V_5 + {^1V_1}$	$^2V_5$	$\left\{ {^1V_5} = {^2V_5} \atop {^1V_1} = {^1V_1} \right\}$	$= 2n + 1$	1	...	...	...	$2n+1$					
÷	$^2V_5 \div {^2V_4}$	$^1V_{11}$	$\left\{ {^2V_5} = {^0V_5} \atop {^2V_4} = {^0V_4} \right\}$	$= \frac{2n-1}{2n+1}$	...	...	...	0	0	...	...	...	...	...
÷	$^1V_{11} \div {^1V_2}$	$^2V_{11}$	$\left\{ {^1V_{11}} = {^2V_{11}} \atop {^1V_2} = {^1V_2} \right\}$	$= \frac{1}{2} \cdot \frac{2n-1}{2n+1}$	...	2	...	...	...	...	...	...	...	...
−	$^0V_{13} - {^2V_{11}}$	$^1V_{13}$	$\left\{ {^2V_{11}} = {^0V_{11}} \atop {^0V_{13}} = {^1V_{13}} \right\}$	$= -\frac{1}{2} \cdot \frac{2n-1}{2n+1} = A_0$	...	...	...	...	...	...	...	...	...	...
−	$^1V_3 - {^1V_1}$	$^1V_{10}$	$\left\{ {^1V_3} = {^1V_3} \atop {^1V_1} = {^1V_1} \right\}$	$= n - 1 (= 3)$	1	...	n	...	...	...	...	...	...	$n - 1$
+	$^1V_2 + {^0V_7}$	$^1V_7$	$\left\{ {^1V_2} = {^1V_2} \atop {^0V_7} = {^1V_7} \right\}$	$= 2 + 0 = 2$	...	2	...	...	...	...	2			
÷	$^1V_6 \div {^1V_7}$	$^3V_{11}$	$\left\{ {^1V_6} = {^1V_6} \atop {^0V_{11}} = {^3V_{11}} \right\}$	$= \frac{2n}{2} = A_1$	...	...	...	...	...	$2n$	2	...	...	...
×	$^1V_{21} \times {^3V_{11}}$	$^1V_{12}$	$\left\{ {^1V_{21}} = {^1V_{21}} \atop {^3V_{11}} = {^3V_{11}} \right\}$	$= B_1 \cdot \frac{2n}{2} = B_1 A_1$	...	...	...	...	...	...	...	...	...	...
+	$^1V_{12} + {^1V_{13}}$	$^2V_{13}$	$\left\{ {^1V_{12}} = {^0V_{12}} \atop {^1V_{13}} = {^2V_{13}} \right\}$	$= -\frac{1}{2} \cdot \frac{2n-1}{2n+1} + B_1 \cdot \frac{2n}{2}$	...	...	...	...	...	...	...	...	...	...
−	$^1V_{10} - {^1V_1}$	$^2V_{10}$	$\left\{ {^1V_{10}} = {^2V_{10}} \atop {^1V_1} = {^1V_1} \right\}$	$= n - 2 (= 2)$	1	...	...	...	...	...	...	...	...	$n - 2$
−	$^1V_6 - {^1V_1}$	$^2V_6$	$\left\{ {^1V_6} = {^2V_6} \atop {^1V_1} = {^1V_1} \right\}$	$= 2n - 1$	1	...	...	...	...	$2n - 1$				
+	$^1V_1 + {^1V_7}$	$^2V_7$	$\left\{ {^1V_1} = {^1V_1} \atop {^1V_7} = {^2V_7} \right\}$	$= 2 + 1 = 3$	1	...	...	...	...	...	3			
÷	$^2V_6 \div {^2V_7}$	$^1V_8$	$\left\{ {^2V_6} = {^2V_6} \atop {^2V_7} = {^2V_7} \right\}$	$= \frac{2n-1}{3}$	...	...	...	...	...	$2n - 1$	3	$\frac{2n-1}{3}$		
×	$^1V_8 \times {^3V_{11}}$	$^4V_{11}$	$\left\{ {^1V_8} = {^0V_8} \atop {^3V_{11}} = {^4V_{11}} \right\}$	$= \frac{2n}{2} \cdot \frac{2n-1}{3}$	...	...	...	...	...	...	...	0	...	...

Fig. 1.6 A program for the "Analytical Machine" by Charles Babbage. From a paper by the Italian engineer Luigi Federico Menabrea from 1842. (Image: Diagram for the Computation of Bernoulli-Numbers, Wikimedia Commons)

Probably more famous in the general public than Babbage is his aristocratic pupil Ada Byron King, Countess of Lovelace (1815–1852). Without any formal mathematical training, she had, as an enthusiastic amateur, further elaborated a programming task by Charles Babbage and documented the associated simple computational program. Babbage probably had similar program lists in his notes, but her program text was published as the "first program" in history (Fig. 1.6). For this purpose, it expands the translation of an Italian article on the Analytical Machine to make it, with its own text (Menabrea 1842). It is a substitute for her own publication, which was probably not possible for her (even as a woman).

The program is tricky, but structurally very simple. It is just a linear flow without "higher" programming elements like branches or subroutines. We single out Ada Lovelace here mainly because of her "loose" and far-sighted remarks.

First, a prescient, serious quote from Babbage in 1869 on the importance of the computer:

"Once the Analytical Machine [i.e., a computer] will exist, it will determine the course of science."

This is certainly correct and has come true. Charles Babbage even foresaw the race of scientists to find the fastest computer or computers, as manifested particularly clearly since 1993 in the "Top500", the annual list of the most powerful 500 computers in the world!

Ada Lovelace goes further in her own writing. In 1842, she foreshadows that numbers are only one kind of multiple objects with which the computer can work. It can also be something quite different, for example, musical notes, if the mutual relationships can only be expressed abstractly. Then the machine could, for example,

"Composing music at any level of complexity or scale."

Even more fanciful, but impressive, in 1844 she dreams of the computer as the "*calculus of the nervous system,*" the origin of thoughts and feelings. Lovelace extends the computer to a general machine, if only on the level of doodles, the casual drawings of stick figures. For "real" creativity, one essential element is still missing: chance. Very poetically she describes it:

"The machine weaves algebraic patterns like the Jacquard loom weaves flowers and leaves."

In weaving, however, the creativity lies with the designer of the pattern, i.e. the human being.

General programmable computers, in the simplest blueprint according to the Hungarian mathematician John von Neumann, have the main components processor (the working part) and a memory that contains the working data and what to do with it (data and the program or software). Babbage finds the idea for storage in mechanical weavers. The French weaver Basile Bouchon uses a storage method for the data of weaving patterns since 1725: Holes in a paper strip or cardboard to store the weaving patterns. Weaver Joseph Marie Jacquard uses punched cards to develop automatic looms in 1804, revolutionizing the textile industry. The American engineer Hermann Hollerith lays the foundation for the company IBM with punched cards and thus for the IT industry, then called "data processing". Figure 1.7 shows the IBM punched card as it was used for program and data input for almost 50 years from 1928.

The first working computer in the sense of Charles Babbage, "Z3", was built as a lonely work by the German engineer Karl Zuse (1910–1995) from telephone relays and was fully functional in 1941. In 1946, he wrote the first high-level programming language, "Plankalkül". Karl Zuse is thus considered the true inventor of the computer, especially of the computer in the sense of a computing aid for the engineer or the administration.

We thus conclude the first, sober part of the history of the computer. There is no uncertainty (the machines, when they work, are error-free) and no chance. We will see

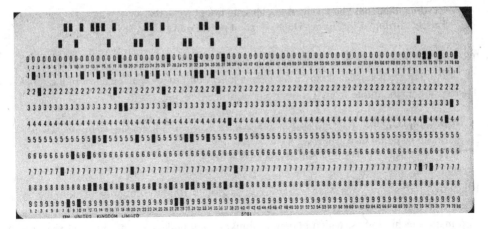

Fig. 1.7 Used (i.e., punched) IBM punch card with 12 rows of 80 columns, circa 1960. (Image: Used punchcard. Wikimedia Commons, Pete Birkinshaw)

the computer below more as a large, very large system and as a philosophical object—then chance and uncertainty will become important again.

1.2.3 The End of the Enlightenment

"The term Enlightenment refers to the development, beginning around the year 1700, of overcoming through rational thought all structures that impede progress."
German Wikipedia article "Aufklärung", drawn on July 2020.

We define in this history of science as the end of the second (of three) epochs of history, that is, the end of the Enlightenment, when the security of immediate Enlightenment (the "reason") ends and no longer reaches further. It is a kind of end of naive innocence. After that, in our language, the time of modernity begins. In physics, time is then unimaginably no longer uniform, but clocks go faster or slower, a particle can be in more than one place, individual atoms are visible, and animal species can change.

But before that is the time of triumph and apparent great certainty: In science everything essential is explored and known, and in technology everything is invented. At least, this is what these famous quotes imply:

In science, he said, everything important has been researched:

"It is never safe to say that physics has no jewels in store like those in the past, but it is yet probable that the great principles have been found. ... A well-known physicist said 'the future truths must be sought behind the sixth decimal'."
Albert Michelson, American physicist, speech in 1894.

And in technology, everything important is invented:

"Everything that could be invented has already been invented."
Falsely attributed to Charles Duell, Head of patent office USA, 1899.

The first quote is confirmed. However, it is often attributed to the physicist Lord Kelvin. The second quote is false and may have been a joke, but it may be in the spirit of the times. There is a genuine quote from a former patent office director:

"The progress of the arts from year to year is hard to grasp. It is possible to foresee when there will be no further progress."
Henry Ellsworth, Head of patent office USA, 1843.

But above classical physics stand as black clouds around 1900 the coming quantum theory, around 1905 the theory of relativity. In addition, a whole area of science is still missing: Information technologies, which belong to thinking and understanding, intelligence, consciousness, sensual feeling, life and inheritance. The success so far is essentially based on the atomic theory, in which atoms are understood as solid spheres, as if they were made of wood or steel, as in the everyday world—how can one explain sensations with this? This still unsolvable task is called the qualia problem. The German physiologist Emil du Bois-Reymond summarized the conflict in 1872 in the motto:

"Ignoramus et ignorabimus" – "We do not know and will not know".

and he asks:

"What conceivable connection is there between movements of certain atoms in my brain on the one hand, and on the other hand, the facts that are original to me, that cannot be further defined, that cannot be denied, 'I feel pain, I feel pleasure, I taste sweetness, I smell the scent of roses, I hear the sound of an organ, I see red ...'"

Figure 1.8 symbolizes the problem and still stands for many philosophical thoughts today: How can I feel red, how can I smell a fragrance, when (exaggerated) I consist only of tiny, moving ("wooden") spheres? A well-known work by the philosopher Thomas Nagel (1974) is entitled: *What is it to be a bat -wie ist es eine Fledermaus zu sein?* He too says that it is impossible to attribute consciousness to physical states. But this is a misunderstanding, the essential thing in the second pillar of the world is not physics, it is only the basis. This is true for any computer. The essential is the algorithm, the systems of algorithms, or in more flexible terms, the "software". So the questions of Du Bois are still valid in 1974:

His original question 5: Where does the conscious sensation in the unconscious nerves?
Original question 6: Where do rational thought and language come from?

Fig. 1.8 A rose. The rose Konrad Henkel. (Image: Wikimedia Commons, Yoko Nekonomania)

The qualia problem will resolve itself as pseudo-philosophical. Du Bois's open questions will be answered in modern times, at least on a philosophical level, or they will simply become obsolete. They will become uninteresting and their terms will no longer be used. New questions will arise instead.

1.3   Modern Science and Technology

"One gets the impression that modern physics is based on assumptions that somehow resemble the smile of a cat that isn't even there."
Attributed to Albert Einstein, but probably made up.
Only common in the German-speaking world.

This original quote—perhaps just well invented, perhaps from the humorous Einstein after all—applies to modern physics, but also to the second pillar of the world: That a computer can recognize a human smile is mystical! But until 1900, physics was tangible. In the previous Enlightenment, physics operated on shock theory with elastic or inelastic solid spheres. You can still do that today, but much of the cosmos, like atoms, has dissolved into the abstract.

1.3.1 Modern Physics and Chance

"There is nothing in the world except empty curved space. Matter, charge, electromagnetism and other fields are only manifestations of the curvature of space."
John Wheeler, American physicist, 1911–2008.

The solid wooden spheres of our normal world actually consist only of empty space with tiny atomic nuclei surrounded by electric fields. All chemistry and all chemical bonds are determined almost exclusively by these electric clouds around the nuclei, also the weaker forces between particles (e.g. the Van-der-Waals-forces). So the world of material things is electromagnetic, in addition there is the well-known gravity: Everything attracts everything by gravity, at earth and at whole universe. Occasionally, radioactivity breaks into our stable world and brings disorder into our clouds of electrons around the atomic nuclei.

The vacuum is empty only at the average. Even in a dark vacuum, virtual particles are continuously created, which can be real for a short time and then disappear again. These energy fluctuations are possible within the framework of the Heisenberg uncertainty principle. As a result, there is a quantum mechanical background noise in the universe. In addition, normal particles also tremble very rapidly and ultra-fine due to a special quantum mechanical effect. This is also internationally called the *"trembling motion"* after Erwin Schrödinger. It is ironically a good German translation for the ancient word "clinamen".

If we add the classical tremors, namely the thermal motion of particles in a gas or a liquid and the thermal oscillations in solids or the thermal fluctuations of electric currents in transistors, tubes or metals, we see: The world is shaky, everything rushes and is, at least on close inspection, uncertain.

Randomness was already present in classical physics, but the usual view was that it was just unimportant perturbation effects that could, in principle, be eliminated. This changes with quantum theory. Now we are dealing with "real" chance, where tracing back to the causes is not possible in principle.

The most famous opponent of "true coincidence" is Albert Einstein, who says this several times:

"[This] brings us no closer to the mystery of the "old man." I am quite firmly convinced that HE does not play dice." Letter to Bohr, 1926.

and

"You believe in a God who plays dice, I believe in law and order and an objectively existing world." Letter to Bohr, 1944.

The allusions in the quotations are pseudo-religious with "the old man", with "HE" and "God". Einstein explicitly emphasizes that he does not believe in any kind of personal

God—but in clear, unlimited causal chains. We define chance more weakly, so Einstein might be more accepting:

▶ **Definition** **A coincidence is a causal chain whose origin cannot be determined.**

The coincidence and the causal chain are, as far as traceable, real existing.

An example is the run of the lottery machine, which delivers a set of numbers. The question (to the machine or its designer) "*Where does this particular set of numbers come from?*" is nonsensical and forbidden. If it were answerable, it would be cheating! But the process is of course causal and real.

It is somewhat ironic to use Albert Einstein, of all people, who does not believe in "real" chance, to explain this definition of chance. We refer to his (mental) elevator experiment (Fig. 1.9): In the elevator, it is not possible to distinguish whether the elevator is standing on the earth or accelerating in space with acceleration due to gravity; it is equivalent. The ball simply falls to the floor of the cabin.

The same applies to deterministic or indeterministic. In Fig. 1.10 the numbers on the right appear "just like that":

If one cannot penetrate the screen for lack of knowledge, it is useless to ask what "really" happens behind the screen of Fig. 1.10. **The effect is identical, whether something is determined or "really" random. The next figure in** Fig. 1.10 **on the right is always a surprise.**

It is certainly humanly unsatisfactory, but acquiring the knowledge simply does not work according to the laws of nature (or the construction of the lottery machine). The emphasis on the fundamental difference between knowing or not knowing a law of formation for a

Fig. 1.9 Einstein's elevator experiment. (Image: Equivalence Principle (excerpt), Wikimedia Commons, Prokaryotic Caspase Homolog)

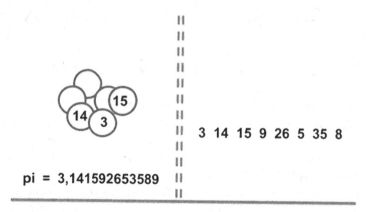

Fig. 1.10 Understanding the concept of randomness. To the right of the screen, the source of the events cannot be identified: Every number is new. On the left is order that can be calculated arbitrarily far

piece of information, whether there is one at all or not, comes from the Argentine-American mathematician and IBM colleague Gregory Chaitin (b. 1947, see Chaitin 2002 and Hehl 2016). Chaitin also states:

> **"The way to randomness is to remove all redundancy. You have to distill it, concentrate it, crystallize it."**

If one knows that behind the screen the number *pi* is generated, everything is determined. In physics, ignorance corresponds to the fact that there is no way to quantitatively calculate the event. The "distilling, concentrating and crystallizing" is just the task of the devices to generate chance, for example, the mentioned lottery machine. Or the physical process is already an elementary process. In the case of a number, it means: there is only the number itself, but no algorithmic prescription to calculate it. The number is simply there. Or in other words: Computable numbers can be represented by a short program, e.g. simply "*1 divided by 3*" or by a continuous recursion, which results in more and more digits in the number system. In the case of a random number, the number itself is the program, possibly a program of unlimited length.

If the "production" of (pseudo-)random numbers is done in the computer, the "distillation" is a problem: It looks quickly quite good "random"; but it becomes difficult with the proverbial "close look". More on the generation of randomness and the problems and possibilities in Sect. 7.3.2.

Quantum theory makes the beginning of the causal chain even more mysterious. At the beginning, there may be entanglement of several objects. An entanglement is a common state of these objects, in the extreme case of the whole universe. The objects cannot be identified individually, only an access to the system provides single information and changes the whole system as a whole! This is, as Einstein said, a "spook", which was proved with elementary particles, for example, photons. For an entangled pair of particles that together form a system, the observation of one particle immediately fixes the second particle as well, regardless of distance. However, this cannot be used to transfer information—the speed of light is a hard limit for the transfer of information.

The well-known thought experiment of the Schrödinger cat drives the thought into the absurd. Figure 1.11 shows the principle of entanglement of objects by the example of a cat, which is in a glass flask together with a radioactive preparation and a poison mechanism, which is triggered by the next occurring radioactive accidental decay (e.g. by a hammer, which smashes a vial with prussic acid, left side of the film). The cat would thus be in an undefined entangled state for that long, simultaneously alive and dead. It is only observation that decides. The film in Fig. 1.11 bifurcates at this point. Either it is alive or dead from now on. Of course, the fictitious example goes too far; the physicist Stephen Hawking didn't even like to hear it anymore *"or else he'll reach for the Colt"*.

The solution is the omnipresent randomness that pelts in on the piston and the cat through the thermal movements of the air, through the incidence of light particles, and much more. This omnipresent noise acts like an observation long before the directed observation and dissolves the entanglement. We will find the effect of the noise again and again and look at it more closely below.

The physicist Stephen Hawking abhorred the beautiful story of the cat as a kind of intellectual popular stupidity, but there is no doubt that entanglement exists in the microscopic world. On the other hand, the doubting Einstein is certainly right when he scoffs:

Fig. 1.11 The Schrödinger's Cat. (Image: Schrödinger's Cat Film Bohn, Wikimedia Commons, Christian Schirm)

"…Do you really think the moon only exists when I look at it?"

No, the moon exists independently of us.

Thus we have a paradigm shift of ideas:

There is effective randomness, (almost) everything rushes, and there are entangled states, even an entangled world background, from which events can arise.

Of course, a little history of twentieth-century physics still included some factual news of "highlights". Here are three essential events:

- 1945: Dropping of the atomic bomb over Hiroshima and beginning of the political atomic age.
- 1964: The discovery of cosmic background radiation as evidence for the Big Bang and the beginning of space, time and matter.
- 1974: Confirmation of the existence of quarks and the standard model of particle physics. It unifies three of the four fundamental forces (electromagnetic, weak and strong interaction) and explains all known elementary particles.

In the course of the second half of the twentieth century, modern physics, especially with semiconductor physics and nanotechnology, has inexorably promoted a technology that is becoming the key technology of the twenty-first century: Information technology. Over 50 years, the number of transistors on a chip has regularly doubled every 1.5 or two years (correspondingly, the performance of computers) and energy consumption has halved. This is roughly equivalent to the famous Moore's Law (doubling of transistors) and the lesser known Koomey's Law (halving of power consumption)—not laws of physics, more journalistic statements about exponential growth. But exponential growth changes everything. This is a well-known phenomenon. It makes a big difference in systems whether you have a few parts, millions of parts, or many, many more particles. It is the difference in physics from the physics of single atoms to nanotechnology or even solid state physics.

It is moving from being a computing powerhouse to key technology, even to understanding the human mind, due to the miniaturization of components and thus the ability to build intelligent systems.

1.3.2 Information Technology and Mind

"I think the original question 'can machines think?' has too little meaning to merit discussion."
Alan Turing, British mathematician, 1950.

The term "think" is too elusive for him, so he introduces the test named after him to avoid a fruitless discussion. He expects that in 50 years it will begin to be difficult to distinguish a human from a machine as an interlocutor. Energetically, he rejects the (then) common theological view that *"only man has been given a soul, and therefore only man, and not an animal or a machine, can think."*

Alan Turing is apparently already in 1950 of the opinion that computers will be able to think so well that the above question would be a pointless argument about words. Today, the question is almost trivial and a kind of tautology: We humans are, after all, also a kind of computer.

Information technology works on a physical basis, but it is fundamentally different from physics. It is the work of and with abstract complex structures that move something physical: Currents in transistors or membranes. IT could also be done with molecules, billiard balls or dominoes, but a modern chip can contain a complex construction with several 10 billion transistors working together reliably. Only nature in the brain can achieve something similar or possibly better—using a completely different technology.

Physics and information belong to two worlds, which the physicist John Wheeler succinctly called the "it world" and the "bit world" (Wheeler also coined the term "black hole"). He even argued in 1990 that information is primary: "it from bit", i.e. that even the material world is actually information "somewhere in its depth". The first to have such a thought was probably Konrad Zuse in 1969, who called it "computing space". Today these are philosophical speculations. We thus emphasize that "it" and "bit" are fundamental pillars of the world, even if for practical purposes the bit world must be built on the "it world". It is "bit from it" without exception.

Two paradigm shifts in modernity through the recognition of hidden information technology emerged early on, still without any understanding of computers:

- The doctrine of the evolution of species by Charles Darwin, 1859,
- The doctrine of the existence of the subconscious by Sigmund Freud, 1916.

Mind, Intellect and Soul Technically Become IT
Probably the most important paradigm shift of modernity is the identification of "mind" and "soul" in humans as the workings of some kind of computer structure with hardware and software, at least from a functional perspective.

A first premonition can be found in the psychologist Sigmund Freud:

> **"[Man] is not even master of his own house, but depends on meager news of what is unconsciously going on in his soul life."**
> **Lecture 1916/1917.**

Sigmund Freud is aware of the paradigm shift in relation to the widespread image of man as an independent, free king, which has penetrated our collective understanding of man from the philosopher Plato via church teachings. Compared to this image of man, the subconscious, discovered by Freud, is a slight. He coins the term Copernicanization:

▶ **Definition** **A paradigm shift that offends humanity is a Copernicanization. Man is literally or metaphorically taken from the center, becomes a marginal figure and more "ordinary".**

According to Plato, man has in himself an immortal, great spiritual being who freely commands man as a whole. The body is almost unimportant and more or less accidental. This is a nonsensical, tragic, though felt by us, separation of body and spirit. Plato was the teacher of the aforementioned Aristotle, who, in contrast, saw body and mind as one and the body as important.

What runs in our mind is technically biological information technology with a chance! The soul in the sense of the psyche is information technology with a lot of randomnesses. We have already proved a lot of mental functionality by reconstruction, e.g. by teaching the computer to understand whether a picture shows a woman or a man, whether the woman is smiling, and so on. Such feelings are recognized, for example, by the program IBM Watson. This program has already won a serious knowledge quiz against humans about 10 years ago (Fig. 1.12). We have listed above the list of human abilities that have meanwhile been proven with IT to be "nothing unique".

Disturbingly, the construction of IT systems seems to have no visible human or natural limits. Thus, according to the author, the sentence applies to the comparison with human functionality:

Fig. 1.12 Scene of jeopardy quiz. Humans lose in knowledge quiz to IBM Watson computer, 2011. (Image: WatsonPour203, Wikimedia Commons, VincentETL)

"Anything a normal human can learn [in terms of skills], a computer [robot] can or could do. And finally this better, more versatile and in combination with things that we humans can't do."
from: "The uncanny acceleration of knowledge", Vdf, 2012.

There is no doubt that the development of IT will continue to affect humans. The supporters of transhumanism, such as the entrepreneur and computer scientist Ray Kurzweil (b. 1948), take an optimistic view of this. We will get used to the "slight" and forget it, but the uncertainty about the future distribution of roles between humans and computers remains.

Biology Becomes Largely IT

"Life is a DNA software system"
Craig Venter, American biologist and entrepreneur, born 1946.

The above quote is perhaps too simplistic when it refers so succinctly only to DNA. But it is certainly correct when referring to the entire web of processes of life: Life is an ongoing software system with a great deal of randomness. Carl von Linné's 1753 and 1758 catalog of plants and animals, respectively, is really the cataloging of types of software programs, the individuals being instances of them. He chose sexuality (e.g., the structure of flowers) as the basic criterion, to the horror of many of his contemporaries; today we recognize sexuality biologically as nature's sophisticated software technique for mixing genes. Linné also catalogued minerals, but there, of course, only physics and chemistry apply.

Without understanding the mechanism even in principle, Charles Darwin had observed the workings of IT around 1838 in the variants of the finches now named after him on various Galapagos Islands and interpreted it correctly: species are not fixed, but are in flux and interact with the environment and other species. The emergence of species is based on evolutionary software technology:

What is needed is the ability to systematically change a program, and a decision criterion, which variant is better in the sense of the task.

The result is evolution, where in nature species adapt to changing circumstances, continue to live, or die out if they fail. Unfortunately, the evolution of species is still called the theory of evolution—but it is not a theory in the sense of a dubious proposition. It is the basis of biology, anchored in physics and geology and astronomy.

"Nothing in biology makes sense except in the light of evolution."
Theodosius Dobzhansky, Russian-American biologist, 1973.

Evolution is generally embedded in an ocean of coincidences of immense proportions, especially in lower organisms. The dependence on coincidences in the 4.5 billion long history of the earth, 3.5 billion of which with forms of life, brings a residue of uncertainty, not into evolution per se, but as far as the underlying philosophy is concerned: Where do the coincidences come from? Are they "entirely" random, or somehow rigged? For the religious, there is as a back door the scientifically neutral idea that a God is behind every coincidence, just as one may think that God had a hand in a car accident. Or only every time there is a good, continuing mutation? Or only in the case of trouble? In this form, the idea of "intelligent design "is neutral, but unnecessary (Hehl 2019). A scientific answer will be possible when one will be able to simulate evolution and see if 4.5 billion years was a good period of time for it, or if the whole thing would have been possible only with some external tutoring!

Evolution continues today, but differently. On the one hand, there is the amplification of human functions through technology, artificial lenses or artificial sensors that can even be implanted. It is also possible to push biological IT further. This is done with synthetic biology, which constructs new biological systems and modifies existing life.

For a brief history of digital information technology, here are a few highlights:

- 1950: Alan Turing proposes the Turing Test; he is convinced that computers think.
- 1953: The structure of DNA is decoded by James Watson and Francis Crick.
- 1957, 1974: The first very large commercial operating systems for computer networks are created, SABRE for airlines, MVS for the IBM/370 computers.
- 1973: Ray Kurzweil builds the first reading machine for general printed type with acoustic output for the blind.
- 2011: The IBM Watson program wins the knowledge quiz against humans.

The fastest supercomputer at the Oak Ridge Laboratory today delivers a computing power of 143 PetaFlops, i.e. 1.4×10^{17} computing operations per second.

The high computational power of digital computers, both of individual chips and of the large systems, also means paradigm shifts: every factor of 1000 gained opens up a new world. A quote says (Northrop 2006):

"Scale changes everything" – "As you get bigger, everything changes".
Linda Northrop, software engineer, 2006.

Depending on performance and program size, a computer can enlarge an image in minutes or milliseconds, can recognize a smile, find a face from hundreds or from millions of images. And the way it works is no less mysterious to most people than understanding how they themselves, as humans, recognize the smile: Billions of crystalline transistors are switched here, quasi-lucid neurons there.

The size of programs brings with it a fundamental, pragmatic problem: a small program is like a machine, a manageable clockwork. A large program contains unpredictability and errors. It contains randomness. As programs run, the operating system creates a network of computational streams that contain randomness, for example, when they are recombined.

To control or reduce the errors in development, an engineering discipline has emerged: software technology. Well known current quotes are

"There's an app for everything." Anonymous,

and

"In short, software eats the world." Marc Andreesen, entrepreneur, 2011.

An error in the program at runtime usually causes the program to crash, sometimes a changed (wrong) result, rarely meaningful new information. In image processing this happened to the author—program errors occasionally gave artistically valuable results. Evolution is based on the ongoing emergence of chance and many program errors.

Thus we recognize a philosophical difference between the world of physics and the world of IT:

Physics describes causal chains and explains the world from the fundamental laws, i.e. "bottom-up".[3] The world of computer science is teleologically oriented and starts from the tasks or "top-down"; one also says "down-causal". Physics works with causal laws, computer science with functional programs and "apps".

Physics explores the limits of the universe; information technology sees no limits yet.

1.3.3 Summary of the Chapter

We have divided history into three periods, ancient, enlightenment and modern, and science into two fundamental areas, physics and information technology.
The definitions of the time limits are clear:

- We define antiquity astronomically: From the early beginnings and as long as circles dominate the world view.
- The Enlightenment: as long as time passes unperturbed, space is flat, and atoms are solid spheres like bowling pins.
- Modernism to the present day with signs of postmodernism.

"Postmodern physics" are for example string theories and multiverses, i.e. concepts without (until today) experimental evidence. Physics is the basis for many other natural sciences, such as chemistry as the science of combining atoms via electron shells, or astronomy as the science of objects and structures in the cosmos. Each of the three sections of the story corresponds to a different attitude towards life.

---

[3] *Bottom-up* and *top-down* generally refer to the direction of action of processes.

IT in the general sense includes all entities with a blueprint, from viruses to humans as biological and intelligent beings, and of course the digital world. In this sense, "postmodern" IT is, for example, quantum computing, transhumanism or digital copies of a human being (the often discussed "mind upload" is probably not a science) or artificial biological beings.

There is a third fundamental pillar of the world besides physics and IT: The chance. Fundamental randomness stems from physics (even if it originates in a human brain), but randomness also operates at the level of IT. We have followed its history closely. Chance is fate, disturbance and foundation of the world.

In Table 1.1 we have tried to summarize on one sheet some of the main players in the short history of science, computer science, and the science of chance. The choice of names is personally coloured and adapted to the aim of the book to put chance in the foreground.

We put a lot of emphasis on paradigm shifts, in science and IT itself, but also in the understanding of history. Ancient science gets excellent marks from us, especially Aristotle, Democritus and Ptolemy. Their deeds or forebodings are great given their beginning without precedents. The most important paradigm shift is the ascent from the mere observation of points of light to the understanding of dynamics, for instance with Newton. Only with this the transition from the geocentric world model to the heliocentric one is accomplished, not with Copernicus and not with Galileo. In addition, there are the beginnings of IT in the form of mechanical machines, for computing as well as for

Table 1.1 Brief history of physics, computer science, and randomness in science

	Ancient	Rec	(Post-)Modernity
Science (physics)	Aristotle Ptolemy Atomism Nicolaus Copernicus Galileo Galilei (Astro)	Galileo Galilei (Phys.) Johannes Kepler Isaac Newton James Maxwell Albert Einstein	Albert Einstein Max Planck Werner Heisenberg Georges Lemaître String theory Multiverse
Computer science	Mechanism of Antikytera	Wilhelm Schickard Joseph jacquard Charles Babbage Ada Lovelace Charles Darwin	Alan Turing John von Neumann Claude Shannon Konrad Zuse Crick&Watson Gregory Chaitin
Random	Clinamen **Chance as fate**	*Opposite of chance: Laplace demon* Statistics Ludwig Boltzmann Henri Poincaré **Chance as a disturbance**	Chaos theory Turbulence Benoit Mandelbrot Uncertainty principle Big data **Chance in the Foundation of the world**

weaving. More possibilities, such as the composing computer, were Ada Lovelace's first dreams.

The Enlightenment ends with the unanswerable question; how can spiritual things be incorporated into this mechanical science? It is an impossible question, because of course it cannot. For that, one must grasp the general principle of the computer as a constructed system. But if one grasps the principle of complex systems, then in modern times one can also grasp Darwin and evolution, and the principle of the human mind and soul. You just have to realize that this is all IT.

IT is not classical materialism, which does not exist in physics any more, but means complex dynamic structure. It is Plato's fault that this is so difficult to accept!

Disturbing things remain: In physics, above all, the existence of real chance, when we also see that "real" chance and ignorance merge. We have tried to show that even classical chance, which has only inscrutable causes, has philosophically the same effect as absolute (quantum) chance.

In addition, there is the physical entanglement of particles (Einstein's "spookiness") and in IT the limitlessness of possible further construction.

Evolution is the development of nature's software up to us.

We humans are software that can construct software ourselves.

The conclusion is a selection of names and terms to the sections of the story. Galileo and Einstein are listed in two eras. Evolution (and thus Charles Darwin, Francis Crick and James Watson) is computer science.

References

Chaitin, Gregory. 2002. *Paradoxes of randomness and the limitations of mathematical reasoning.* New York: Wiley Onlinelibrary.

Hehl, Walter. 2016. *Wechselwirkung – wie Prinzipien der Software die Philosophie verändern.* Heidelberg: Springer.

———. 2018. *Galileo Galilei kontrovers.* Heidelberg: Springer.

———. 2019. *Gott kontrovers.* Zürich: Vdf.

Menabrea, Luigi Federico. 1842. Sketch of the analytical engine invented by Charles Babbage.en. wikisource.org/wiki/Scientific_Memoirs/3/Sketch. Zugegriffen im Sept. 2020.

Northrop, Linda. 2006. *Ultra large systems. The software challenge of the future.* Pittsburgh: Carnegie-Mellon.

Schlichting, Joachim. 1993. Physik- zwischen Zufall und Notwendigkeit. *Praxis der Naturwissenschaften, Physik* 42/1: 35.

Zekl, Hans Günter. 1986. *Aristoteles Physik: Vorlesungen über Natur.* Hamburg: Meiner.

The Coincidence Itself

2

"Random chance is not a sufficient explanation for the universe - in fact, chance cannot even explain chance."
Robert Heinlin, science fiction writer,
in 'Stranger in a Strange World', 1961.

Coincidence, by definition, explains nothing: We speak of chance precisely because we have no causal explanation. So it is also a fallacy to want to explain free will with the action of chance in the brain. This would mean that a decision would not be made by the homunculus in us, the "free" I, but by an abstract dice game that runs in the brain. That would certainly not be a free decision of the ego. More on this below.

2.1   The Right Word

"A well-chosen word can save a tremendous amount of thought."
Ernst Mach, Austrian physicist, 1838–1916.

The word itself is a problem, but less so in German. The German word "Zufall" meaning "*falling in*" shows very well how an event comes in from the outside, from the unknown, in contrast to the philosophical necessity of an event whose cause can be traced back.

In English, we speak of *random* and *randomness*, of *chance, accident, hazard* or, more elegantly, of *contingency*.

According to Wiktionary, *"random"* is related to the German *"rennen"* and the French *"randonner"*, roughly in the sense of *"wandering"*, but probably aimlessly and with a lack of a specific direction. "Chance" and "accident" generally carry an evaluation:

Chance (it comes from the fall of the dice, from the Latin "cadentia"). Chance tends to "chance favorable", to be lucky, conversely an *accident* is rather an unfavorable event that

© Springer Fachmedien Wiesbaden GmbH, part of Springer Nature 2021   35
W. Hehl, *Chance in Physics, Computer Science and Philosophy*, Die blaue Stunde
der Informatik, https://doi.org/10.1007/978-3-658-35112-0_2

Fig. 2.1 Dice as symbols of chance. (Image: A_throw_of_two_dice_as_ in_the_game, Wikimedia Commons, Jon Richfield)

falls in. Add to this the word branch "*casualty*", also from Latin casus ("fall") and cadere (to fall), which has also slipped into negative meaning: In English, it also directly means "military casualties," i.e., fallen soldiers.

In philosophy, the term *contingency*, from the Latin *touch,* is used today in the sense of unpredictability, of something that may or may not happen.

The word hazard, common in many languages, probably comes from the Arabic word for dice *az-zahr.* The dice (Fig. 2.1) are symbols of chance with secondary meanings ranging from chance with luck to adventure to danger. The human question of fate is then obvious.

How essential language is for philosophical understanding is shown by the popular English definition for coincidence, called the Commonplace Thesis ("thesis in everyday language") according to the Stanford Encyclopedia of Philosophy:

"Something is random iff it happens by chance".

"Something is accidental if, and only if, it happens as an accident".

The logical "iff" expresses the unconditional "if". The translation to German creates a tautology. In English, it may seem less tautological, but the problem does not disappear!

The German word *Zufall* fits perfectly. It is a medieval translation of the Latin word *Accidentia* and expresses that it is a process which is not only disordered, but it looks as if "it comes from outside", that is, from outside of real, tangible nature. The question of "whence" cannot be answered. The word is emotionally neutral and has neither positive nor negative connotations. It is a stroke of luck (perhaps a happy coincidence?) for German culture compared to the otherwise dominant English. What buzzwords should one use when searching English for the "science of randomness" in the sense of this book? "Randomness" is too colorless, "fate" too strong. We will use the term *random* only when talking about undifferentiated, initially harmless randomness.

I would like to suggest to the English language community that the word *Zufall* be adopted as a loanword, like the words *kindergarten, gestalt, edelweiss,* and *leitmotif.*

In ancient Greek thought, several expressions describe "chance". The best known is *Tyche*, originally the personal goddess of fate, who brings the (unforeseen) will of the gods to people and changes their fate (see Fig. 8.6 and under "Tychism"). The origin is the verb *tychanein* meaning "to hit a target with a projectile," then "to be lucky" and "to succeed." (Mauthner 1917/reprint 1997). It is chance that works or can work in human actions. Aristotle's example of *tyche* is someone who exceptionally goes to the marketplace and then just "by chance" meets his debtor and receives his money.

The coincidence in nature is called *automaton*, for example, when a tile falls from the roof in a storm, unless someone has loosened the tile. Then it would not be a coincidence.

Another term is *symbebekos* in the original meaning of "coming together". It is the coincidence that occurs, for example, as a by-product in a planned action, so *symbebekos* is about:

> **When someone finds treasure while digging a hole for a plant.**
> **When someone arrives at a different, unplanned location on a trip.**
> **Aristotle, Metaphysics V,30.**

There is a confusing literature and history on the interpretation and containment of these ancient Greek terms in the environment of "coincidence" and the resulting Latin loanwords.

2.2   Coincidence and Necessity: Introduction

> **"Everything that exists in the universe is the fruit of chance and necessity."**
> **Democritus, Greek philosopher, 459 B.C.- 370 B.C.**

This is a famous quotation, but its authenticity is disputed. The expression is too modern and no ancient source can be found. It was disseminated in this form by the French biologist and Nobel Prize winner Jacques Monod, (1910–1976). Monod made the quote the title of a successful book on the philosophical significance of evolution (Monod 1970). But it is also a possible motto for the present book.

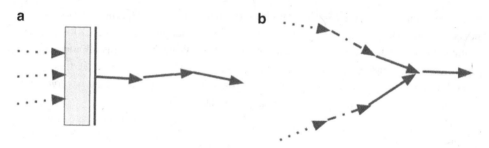

Fig. 2.2 Illustrations of the definition of randomness. The time axis goes to the right. (**a**) The wall is impenetrable to the left. Randomness appears from the surface. (**b**) At least one of the causal chains is also blocked to the left

2.2.1 Definition of Randomness

We define chance as a causal chain that starts without a past (as is the case with dice by definition) or as the meeting of two causal chains that, as far as they can be traced, have no connection (Fig. 2.2). It is common to call chance an "event" in this sense.

> **This means that you can understand a coincidence in a general context, but not in a specific case.**

One understands causality in principle, but not in detail. In the case of dice, one understands the picture of the cube, the momentum and angular momentum theorems, but the process of dice itself is obscure. The general context is given by physics and its laws that are followed. "Necessity" is the philosophical word for the "causal preconditions" without which the process under consideration does not occur.

It is not easy for people to think "a causal chain begins", a process that has no past. "It" has to go back further and further, cause by cause. Causal thinking is given to us from evolution, and that makes sense. The processes we devise to destroy the past (shuffling cards, rolling dice) and the machines we build to do it (roulette, lottery machines) remain causal, only the processes are so convoluted and dependent on such subtleties that the wall in Fig. 2.2a is impenetrable in the direction of "past". The wall is the totality of the entangled inner processes in the mechanism. There is no information above the wall and to the left of the wall. A nice but not very meaningful philosophical description of the situation comes from Aristotle and Hegel: "*Chance is its own cause.*" (Kaiser 1990).

For the dice roll, the wall blocking the view back is a good comparison. In the sense of the general philosophy of chance, a noisy, structureless cloud as the origin of chance is a better image (see section "Noise").

It is hard to accept that there should also be "real" coincidence or "real" Zufall, but it is confirmed in the world of quantum physics: The radioactive decay of a single atomic nucleus cannot be predicted from the available information, only with a large number of atoms a well measurable distribution results. Something happens without reason, but still in the aggregate according to strict laws. At the atomic level we have wave functions and superimposed states. If they decay, in our macroscopic world an event, for example, an elementary particle, appears as a real coincidence!

2.2.2 Causal Chains

A single causal chain, e.g., when rolling a die (Fig. 2.2a), is an idealization as well as the meeting of two causal chains with the generation of a new event (Fig. 2.2b). We are accustomed, for example, to regard a collision of two cars or the crossing of a pedestrian on the crosswalk as a coincidence, as the intersection of two causal chains—unless in a criminal case, when it would be premeditated murder. When we speak of a coincidence, we assume that at least one of the causal chains is lost in the unknown. Figure 2.3 shows two cars and the perplexed drivers who mystically met at this intersection!

In this view, a person's life is a series of coincidences and their management.

This was precisely the point of view of the existentialist conception of life. According to Jean-Paul Sartre, life is shaped by many unpredictable coincidences (see below). From the point of view of social relations, people's lives are a continuation in the common network of coincidences. This is the stuff of novels and the subject of world history! This is where the incredible events take place, which touch us as the "power of fate".

On this psychology of feeling *this is unbelievable* in random events, the English mathematician John Edensor Littlewood (1885–1977) made a mathematical consideration,

Fig. 2.3 Meeting point of two causal chains. A car accident as an example. (Image: Japanese_car_accident, Wikimedia Commons, Shuets Udono)

published by the physicist Freeman Dyson. They define an event as "incredible" if it has a one in a million chance of occurring. To do this, he assumes that we are capable of registering an event every second, whether it is striking or within normal limits. This gives, with eight hours of full attention in 35 days, a million possible events. So the individual experiences something "miraculous" in this sense about once a month (Littlewood's "law"). Extending the view to the coincidences that groups of people experience, especially to humanity today connected by the media, we can say:

> **With a large enough sample, any outrageous thing is likely to happen.**
> **Law of truly large numbers, according to mathworld.wolfram.com.**

The purpose of these reflections or "laws" is to warn. Some events are unbelievable, but not supernatural. It is a warning against jumping to conclusions: Telepathy does not exist, nor does the synchronization of the dream world and the real world, as CG Jung supposed. But there are very improbable processes. This message reached me at the moment of writing about Littlewood's Law:

> **"A man wins a $4 million lottery jackpot - for the second time"**
> **CNN News. 06/27/2020.**

An incredible coincidence: The double win and the matching message for me!

It is characteristic of chance events that they produce something new, but still natural, such as the appearance of an elementary particle or a black swan, or at least a discontinuous change of direction when two molecules collide in Brownian motion. Thus chance is at odds with the way we think. We think in terms of continuous changes in the real world (Fig. 2.4). In classical mechanics, bodies move on steady, smooth curves, even curves that have minimal curvature and run as "straight" as possible. This is the Hertzian principle of mechanics.

The idea of continuity in nature can already be found in Aristotle and runs through antiquity and the Enlightenment. The concise formulation comes from the biologist Carl von Linné in 1751:

> **Natura non facit saltus - Nature does not make leaps.**

The epitome of causality and continuity is the description of events in the world in the form of differential equations. Here one takes advantage of the fact that one can often formulate the relevant quantities at a point in time and at a point in space, including their changes and the speed of their changes. From this one can calculate the future at least numerically.

Figure 2.4 illustrates forms of continuity and types of causality. Sketched are trajectories of a fictitious object in space as a function of time. Figure (a) demonstrates a world in which similar initial conditions produce similar trajectories. This is called strong causality. The

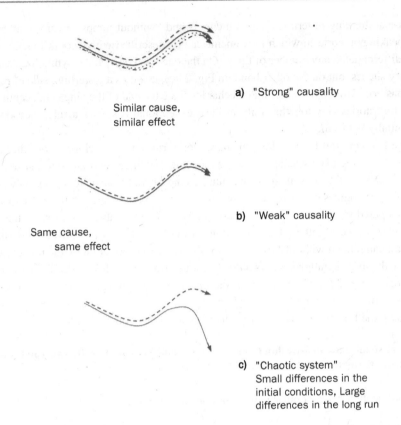

a) "Strong" causality

Similar cause,
similar effect

b) "Weak" causality

Same cause,
same effect

c) "Chaotic system"
Small differences in the
initial conditions, Large
differences in the long run

Fig. 2.4 Illustrations of the continuity of the classical world. The time axis goes to the right

bundle of trajectories with small initial changes essentially stays together. Figure (b) is intended to represent a situation where only a repetition with *perfectly identical* initial conditions produces the same trajectory. This is "weak causality." The requirement of identical initial conditions is illusory in reality even in classical physics, it is metaphysics. The ancient, pre-Socratic philosopher Heraclitus of Ephesus, sensing this, captured it with his saying *panta rhei* (everything flows): *"We get into the same river and yet not into the same one"*.

In the quantum world, it is explicitly impossible to produce identical states because of the uncertainty principle. The two concepts, strong and weak causality, are more philosophical concepts than physical ones.

In sketch c) there is a small change in initial conditions in location or initial velocity between the two curves and the orbits drift apart; it is a chaotic evolution. After a sufficiently long time, the relationship of the two bodies to each other looks independent and "random".

Classical thinking in terms of smooth curves and "without jumps" means, mathematically speaking, a world in which movements can be described with smooth functions that can be differentiated any number of times. On the one hand, sketch 2.4 symbolizes smooth trajectory shapes, but on the other hand, in Fig. 2.4c), it shows a disturbing side of reality, which has been known from celestial mechanics since the end of the nineteenth century: In reality, trajectories stay together only in ideal exceptional cases; as a rule, they diverge exponentially with time.

While two celestial bodies alone in space would run on stable ellipses (or other conic sections) according to classical physics, the movement of three celestial bodies is generally no longer stable and there is no general solution. One can only calculate numerically on the computer the progress of the bodies. If we consider the case where the third mass is very small compared to the other two, there are stable (or nearly stable) positions of the three bodies relative to each other. These positions are important for space travel. Only there a spacecraft can remain without correction or only with little correction by the rocket motor. These are the five Lagrangian points named after the astronomer Joseph-Louis de Lagrange (1736–1813) (Fig. 2.5). All other configurations are unstable and go through a wide variety of orbits, some of which look very random.

In 1899, the French physicist and mathematician Henri Poincaré stated:

"A very small cause that we don't notice causes a considerable effect that we can't miss, and then we say the effect is random."

This amplification has a popular name today: Butterfly Effect.

2.2.3 Understanding with Chance: Astronomical

"The man in the moon always looks down on the earth with the same face. "But the other side of his face is hidden from our gaze. Why?"
Question in Science.com, 2017.

Fig. 2.5 Lagrange's points in the three-body problem of astronomy. (Image: Lagrange_very_massive, Wikimedia Commons, Inkscape, EnEdc)

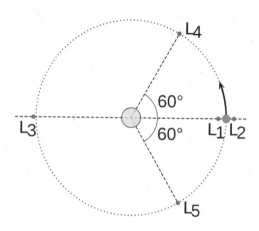

There are some pretty cases of coincidence in astronomy, in our solar system. I do not necessarily mean the structure of the solar system itself, such as the distances of the planets from the sun.

Kepler sought an order here by nesting the regular solids one inside the other in such a way that the sphere inside one shape coincided with the outer sphere around the next shape. He had chosen the octahedron (the 8-flat) with Mercury inside, Venus outside, then the icosahedron (20-flat) with Earth outside, the dodecahedron (12-flat) with Mars outside, the tetrahedron (4-flat) with Jupiter and then the hexahedron (6-flat) with Saturn on the outer sphere. Wonderful, mystical but without meaning (a bit of a paradox, no?). The coincidence Kepler found is almost coincidental. Kepler was thrilled, he thought he had found the divine order:

"Nowhere is there anything too much, nowhere is there anything too little; nowhere is there a point of attack for criticism."

But his ideal order does not exist, it was illusion. There is, however, a certain regularity in the distances, blurred with chance. It is the "law" of Bode-Titius, an empirical formula for the distances of the planets from the sun

$$a_n = 0.4 + 0.3 \times 2^n$$

measured in astronomical units and compared with appropriate values of n:

$-\infty$ for Mercury (i.e., the second term is omitted for Mercury) and then n $= 0, 1, 2, 3, 4, 5, 6$.

for the other planets from Venus to Uranus, where n $= 3$ falls on the planetoid Ceres. On the one hand, the rule originates from the formation process of the solar system and, on the other hand, is determined by ingrained resonances in the system that suppress chance:

Resonance occurs when two (or more) planets meet again and again after integer numbers of repetitions of orbits.

So it has to be as accurate as possible:

n orbits planet A equals m orbits planet B.

The smaller the numbers n and m, the more stable the resonance. We discuss two examples that show how rules can emerge in randomness.

Earth, Venus and Coincidence

If, as a human observer, one follows the course of the planets over the years, it looks as if all of them, including the Earth, run their courses unaffected and meet each other occasionally by chance. These encounters are called conjunctions and oppositions and, as mystical events, are the pseudo-basis of astrology. If you plot the position of Venus over the years as seen from earth, you should get a sequence of overlapping loops that just go on and on,

Fig. 2.6 The apparent magnitudes of the Sun and Moon, minima and maxima. On the left the moon, on the right the sun. (Image: German Wikipedia, Tdadamemd)

randomly tangling. Figure 1.4 in Chap. 1, shows what can really be seen: After eight years, the result is a harmonious rosette, a pentagram![1]

The picture is far away from coincidence. The reason is that into the general coincidence a resonance breaks in as "almost-order" (note who breaks in!). To the eight earth years correspond 13.004 venus years. In a resonance, two celestial bodies get a little hooked up, so to speak, and pull a little toward each other. Nevertheless, over thousand years one sees only coincidence! In the solar system there are quite a few such resonances, also with several celestial bodies together. The best-known resonance, however, is in the Earth-Moon system.

Earth and Moon and Coincidence

Two coincidences are generally known of the Moon: The Moon has about the same apparent size in the sky as the sun, namely about 30' (Fig. 2.6), and the Moon always shows us the same side, or more precisely a little more than one half, due to the leading or lagging on its uneven orbit, the so-called libration (Fig. 2.7).

---

[1] The picture has nothing to do with historical epicycles.

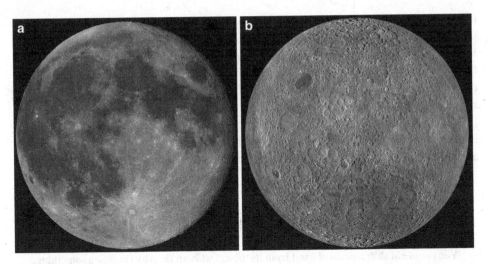

Fig. 2.7 The two halves of the moon. (**a**) The side facing the Earth. (Image: Wikimedia Commons, Luc Viatour). (**b**) The side permanently facing away from Earth. (Image: Moon Farside LRO, Wikimedia Commons, NASA)

Without knowledge, both appear to be coincidences, but the explanation for "always showing the same side" is, as above, a resonance: The moon also rotates around itself once in a month during its orbit around the earth. Earth and Moon are gravitationally locked into each other via their internal imbalances. The tides slow down the Earth's rotation, by about 2 milliseconds per century, and they lift the Moon nearly 4 inches farther from the Earth each year. The day on Earth will thus get longer, and it will eventually match the length of the month, which will also get longer. The day and month will be the same, lasting about 47 days, and the Earth and Moon will both show the same sides to each other.

Let us observe from these examples when we understand something (causally):

- *"Sun and moon have same apparent size"* remains a "coincidence". It is a synonym for non-understanding. We do not find a clarifying relationship.
- *"The moon always assigns us the same side"*. This we understand. The one understood relationship of the two objects to each other in the otherwise unknown is sufficient for this.

2.2.4 Chaotic

"If the flap of a butterfly's wings can help trigger a tornado, it can help prevent one."
Edward Lorenz, American meteorologist, 1914–2008.

The mathematical meteorologist had simulated weather on the computer in 1961, still on a computer with electron tubes. In one run, instead of his exact desired input value, he gave this number rounded down, as he had it on a printout just at hand: Instead of 0.506127, only 0.506. But the result of this simulation looked completely different and the two weather forecasts were completely different. That little change had totally altered everything. At first he commented on his guess as "*the flap of a seagull's wings can change the weather*", then it became the more dramatic cause-and-effect combination "*butterfly in Brazil is responsible for a tornado in Texas*".

It is important to understand that we are not talking here about individual critical points in the system that can tip over (see the Norton Dome below). The butterfly effect expresses a general property of the world with its interactions. Already Aristotle writes in this sense:

"The slightest initial deviation from the truth is later multiplied a thousand fold",

and the German philosopher Gottlieb Fichte (1762–1814):

"You could not shift a grain of sand from its 'place' without thereby changing something, perhaps invisible to your eyes, through all the parts of the immeasurable whole."

Aristotle probably means it quite soberly, with Fichte it is the expression of a holistic romantic feeling for the world. Finally, the astronomer Pierre-Simon Marquis de Laplace (1749–1827), famous for his answer to Napoléon Bonaparte why he did not mention God in his work, took sobriety to the extreme:

"*Citoyen premier Consul, je n'ai pas besoin de cette hypothèse*" -Citizen and First Consul, I have not needed this hypothesis.

The above quotation is probably incorrect in this general sense and is meant quite differently by Laplace (Faye 1884). Napoleon had probably pushed the understanding of the solar system as a clockwork so far that he thought God had to wind up the heavenly clockwork again and again. This was denied by Laplace: God does not intervene in the world and does not need to.[2] On the subject of *God*, see, for example, Hehl (God Controversial 2019).

Laplace is sure that an intelligence (today called Laplace's demon or Laplace's ghost) with a world formula (in today's language) and the knowledge of the situation of the world would be capable of

"To comprehend the movements of the largest celestial bodies and those of the lightest atom. Nothing would be uncertain to them [the intelligence], future and past would be clear before their eyes."
Pierre-Simon Laplace, Essai philosophique sur les probabilités, 1814.

---

[2] For a discussion of the quote, see, for example, the English Wikipedia article Pierre-Simon Laplace.

This statement is the pinnacle of supposed certainty in the Enlightenment worldview of science: Everything is determined, there is no such thing as chance. By the end of the nineteenth century, Poincaré suspects that this cannot be true, that Fig. 2.4c applies to physical processes, and that the life trajectories of two similarly launched points drift exponentially apart in their orbits.

Laplace was far too optimistic. There are fundamental limits to the calculability of the world. Already classically, we can only approximately determine the initial conditions, but even more so in quantum theory. A mathematical measure for the divergence is the Lyapunov exponent after the Russian mathematician Alexander Lyapunov (1857–1918). It describes the speed with which two initially neighbouring points with almost the same initial speed move away from each other. With a positive exponent, the distance grows exponentially and thus inexorably.

It follows that even our solar system, the epitome of stability, cannot exist forever, even if the sun remained the same. (It expands on the one hand by burning hydrogen, on the other hand, it loses mass by the solar wind). The order in the solar system is disturbed by the interactions of the planets and asteroids with the sun and with each other and additionally by further effects, for example, the imbalances in the interior of the sun, its quadrupole moment. The first thing to shift is the position of the Earth in its orbit, then the orbital elements such as the inclination and shape of the ellipse will fluctuate. The solar system is chaotic on longer times. Lyapunov times[3] for planetary orbits are about 5–20 million years (Sussman and Wisdom 1992)—a short time compared to the existence of the solar system. As a result, different components and elements of the solar system become chaotic and unpredictable at different rates. Only resonances occurring in between can stabilize motions for a while. At orbital elements, it is the locations of celestial bodies at their orbits, which shift first, at planets as a whole the orbit of Pluto is most sensitive.

It is typical for the reality and complexity in the world that many parts like the sun and the planets interact with each other; physics calls this in mechanics the "N-body problem" or in quantum theory the many-body problem. There is no exact solution to this. The only solution is ultimately the course of the world itself.

2.2.5 From Atomic Coincidence to Great Effect

"Give me a fixed point and I'll unhinge the world."
Archimedes, Greek mathematician and philosopher, 287-212 BC.

The atomic world, i.e., "at the bottom", is from the very beginning the world of random jumps: Light, for example, consists of jumps. We have already mentioned Brownian motion in the light microscope at the boundary between atomic scale and the everyday

---

[3] It is the time span for drifting apart by a factor of e, Euler's number 2.71828.

Fig. 2.8 The trajectory of a
particle in a Brownian motion.
(Image:
Csm_Brownian_Motion,
Wikimedia Commons,
NivedRajeev)

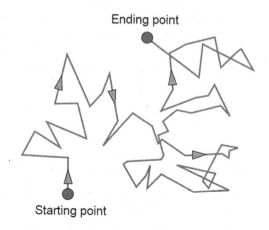

world. The sketch of Fig. 2.8 shows the discontinuous character of the motion; the curve is not differentiable at the kink points. It demonstrates the randomness and disorder of the atomic world. The botanist Robert Brown had at first thought the teeming motion to be the sexual life of plants. The conservation laws of energy, momentum, angular momentum, and electric charge, which are also observed in the trembling motion, act as ordering principles. These theorems are deeply anchored in the world by internal symmetries according to the theorem of the German mathematician Emmy Noether (1882–1935).

However, the thermal tremors of atoms are small, e.g., the mean free path of a molecule in air is 68 nanometers, which is one-thousandth of the thickness of a human hair. One liter of air contains 2.8×10^{21} molecules. These very large numbers or, conversely, the smallness of atoms or molecules in the order of nanometers (in the case of the water molecule, about 0.3 nm) ensure that we are mostly unaware of the fluctuations in the atomic world. Even the visible tremors under the microscope (Fig. 2.8) are actually rare events and the tip of the iceberg: Among many millions of impacts on the particle, one impact of a water molecule is then strong enough to visibly shake the lycopod spore or the ink particle.

But it is quite easy to use this knowledge to build a device that catapults the collision of a water molecule into the macroscopic world. Let the device for observing Brownian motion be given: Optical microscope, microscope slide with water droplets including particles, a stopwatch, a switch, detonator, and a bomb (Fig. 2.9). *The observer selects a particle in the image to observe and starts the stopwatch. After 10 seconds the clock is stopped and a decision is made:*

If the particle has moved significantly upwards, the bomb is triggered, if fuzzy or down, don't.

The experiment is classically and physically problem-free, a kind of God's judgement. It provides a philosophical and physical link from the atomic to the almost arbitrarily large—it could even be an atomic bomb.

Image Microscope:
Olympus CH2 Microscope2,
Wikimedia Commons, Amada44

Image Brownian particle motion:
After Csm_Brownian_Motion,
Wikimedia Commons, NivedRajeev.

Image Bomb:
Mark 7 Nuclear Bombat USAF Museum,
Wikimedia Commons, Chairboy.

Fig. 2.9 Illustration of the thought experiment "Linking the atomic world with the macroscopic world"

It is an example in the spirit of the German bio- and physical chemist Manfred Eigen (1927–2019), who wrote in 1975 (Eigen and Winkler 1975):

"Chance has its origin in the indeterminacy of these elementary events. [. . .] Under special conditions, however, it can also come to a build-up of the elementary processes and thus to a macroscopic mapping of the indeterminacy of the microscopic dice game."

In the butterfly effect, we described a purely physical amplification of a small, random cause. The thought experiment amplifies a minor physical coincidence to the macroscopic level. This is proof in principle that it is possible and that not all atomic effects are or need to be imperceptible. The simplest and most powerful way to amplify an incidental effect is through the means of information technology. We see this in the viral spread of an attractive message on the Internet. This is exactly what happened biologically in the evolution of life. With information technology in the form of life and with us humans, nature has indeed unhinged the world. At the beginning and during evolution, tiny molecular coincidences repeatedly decided further development.

2.2.6   From Single Random Event to Statistical Law

"They should call it entropy, because nobody really knows what that is."
Proposal for the designation of John von Neumann to Claude Shannon for his theory of information.

From the sum of the set of individual coincidences, each apparently arbitrary in itself, astonishingly laws emerge. The classic example is the description of gases as vast numbers of colliding molecules. The individual collisions are not (or only with difficulty) observable, but the effect of the many produces the macroscopic properties of the gas, such as the pressure, the viscosity, the thermal conductivity, the specific heat.

The Normal Distribution

The visibility of laws starts already with a small number of coincidences in an experiment. A particularly impressive example is the Galton board, an experimental realization and visualization of the central limit theorem (Fig. 2.10).

Sir Francis Galton (1822–1911) was a versatile British naturalist, among other things pioneer of dactyloscopy and inventor of the dog whistle or Galton whistle. The dog whistle produces particularly high-frequency sounds that humans can no longer hear. In statistics, for example, he coined the term "regression to the mean". This is a cognitive illusion that occurs when observing stochastic events. If one measurement is extremely off, the next events will likely be closer to the mean. This effect is intuitively difficult to understand and often leads to erroneous conclusions.

The Galton board consists of an arrangement of pins as obstacles for falling balls. The basic cell of obstacles in the set of pins consists of five pins in the arrangement of the five points on the die face for "five", called a quincunx. The balls fall on the center pin and on to the left and right with equal probability. Watching the seemingly chaotic fall of balls through the board, one observes the systematic structure of the statistical distribution, the normal distribution as a bell curve. Francis Galton was fascinated by the falling balls:

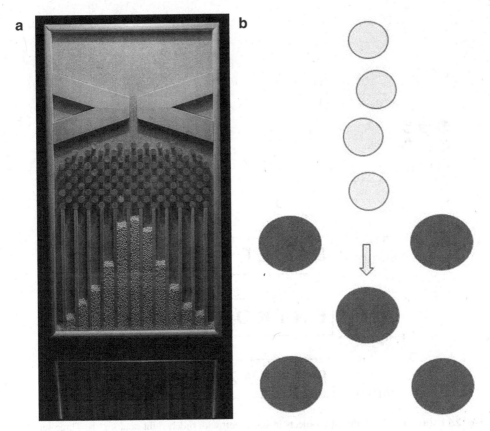

Fig. 2.10 (**a**) A Galton board for visualizing the normal distribution. (Image: Bean machine, Wikimedia Commons, Rodrigo Argenton). (**b**) A quincunx, the basic unit of the Galton board for the scattering of the spheres

> **"I know of few things that can impress as much as this wonderful form of cosmic order [...] it turns out that an unexpected and magnificent regularity is hidden everywhere."**

Laws are hidden even in chance. The image of the running Galton board (Fig. 2.11) also directly visualizes the concept of physical "frequentist" probability. It is, in each case for one column, the number of balls in a column divided by the number of all balls dropped in the experiment up to that point.

In the teachings of statistics, chance is hidden and averaged out. The word statistics itself comes from the Latin *statisticum "concerning the state"*. The properties of the individual are much more random than the statistical values of the totality, such as those of the state. This is well expressed by an old word for statistics: *Collective research*. Thus, in the nineteenth century, the individual disappeared in the aggregate and became a proverbial number. Only in the twenty-first century, with the great capacities of computers,

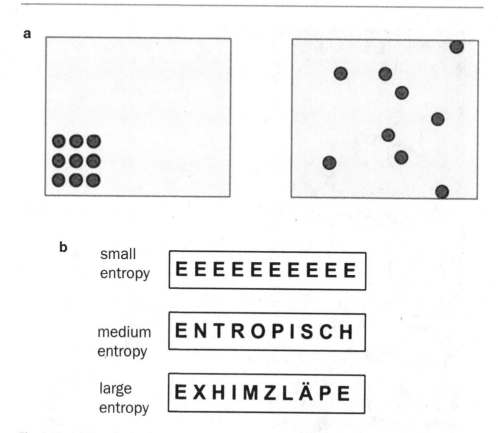

Fig. 2.11 (**a**) Two distributions of objects in space, left with order, right random. (**b**) Three sets of letters with different entropy Above the smallest, below the highest entropy

is it possible to see the aggregate as a set of individual random data and analyze it down to the individual. This is the information technology of Big Data.

Entropy in Physics

Since its introduction by the German physicist Rudolf Clausius in 1865, the concept of entropy has been regarded as "difficult to understand, spooky, meaningless, very special".

It began with the neologism, a made-up word from the Greek ἐν 'an', in English 'inside' together with τροπή 'turn'.

Clausius is probably alluding to the natural direction in which entropy grows:

> **"The energy of the world is constant. The entropy of the world tends towards a maximum."**

Difficult to understand is indeed the abstract formulation of entropy in thermodynamics, but the concept becomes more vivid when you see the atoms themselves.

The concept of entropy in the atomistic sense and the sense of information is in principle easy to grasp. Entropy, according to the Austrian physicist and philosopher Ludwig Boltzmann (1844–1906), is a measure of the degree of disorder (or order) in a system or, in the case of information, a measure of the information required to acquire and generate a set of data. The formulation is deliberately chosen so that, in the sense of the theorem above, entropy is defined as increasing. Thus it holds:

Physically: the more disorder, the higher the entropy, and for information: the more complicated a word or text, the higher the entropy.

Disorder on the one hand and lack of information on the other are both forms of circumscribing coincidence, directly or indirectly!

A common popular form of the entropy law *"entropy always increases by itself"* is the experience that for some inexplicable reason individual socks are always lost in the washing machine, reducing the order of pairs of socks. So it holds:

"The number of complete pairs of socks is constantly decreasing, i.e., the entropy of the system of socks is increasing".

The two sketches of Fig. 2.11a) illustrate the physical form of entropy as a measure of the non-uniformity with which space is exploited. The order (left) can be described compactly and briefly, the random arrangement (right) only with the coordinates of each point, thus with much more information. Of course, the locations of the uncountable number of trembling or oscillating molecules in a true gas cannot be explicitly described; macroscopically, information about all particles must be replaced by a blanket measure of disorder, and thus of ignorance and inherent randomness. This is the thermodynamic entropy. For absolute zero, the entropy is set to 0. From this, the entropy of a quantity of matter can be measured (in units of energy/temperature, joules/kelvin). The absorbed heat of a body is related to its internal disorder: It wanders into all possible degrees of freedom of the molecules and generates movements, oscillations and rotations.

Entropy in Information

Entropy for sets of information, e.g., of texts, is to be understood accordingly as a measure for ignorance and coincidence. This is demonstrated by the three examples in Fig. 2.11b. The lowest entropy has the sequence of equal letters; one could abbreviate it to "10xE". The next level of entropy content has a familiar word. You can guess the whole text even with fragments. For example, in German, "sc" is likely followed by "h"; such information reduces randomness. The highest degree of disorder is in random text, where all characters are independent of the others. Each character must be encoded individually; the larger the choices (e.g., only uppercase letters, or uppercase and lowercase, or even with digits and special characters), the greater the overhead. When creating passwords, one just follows the goal of the highest disorder.

With the concept of entropy, the atomic or molecular chance is hidden everywhere in the world. Furthermore, we have seen from the text example that randomness is closely related to disorder and ignorance. For a system with high entropy, a lot of information is needed to describe it, most of all for a completely random text. In other words, a (truly) random text cannot be compressed. The statement of the mathematician Gregory Chaitin (1969), though better known under the name of the Russian mathematician Andrei Kolmogorov (1965), is valid. It is actually the algorithmic definition of randomness:

With a random object, the object itself is its encoding. You can't get more compact than that.

The Complexity of Software

Just now we constructed texts with little or more effort depending on the internal clutter. We now generalize the construction process instead of dead letters to commands, e.g., for the computer. This is a generalization because, after all, a command could be "write the letter E". They are generally commands to the computer to do something. In the totality of the commands, this becomes a program. The programs for the output of the letter sequences of Fig. 2.11b) are naturally very short. The longest program in the example corresponds to the random letters—it takes the string character by character and simply outputs it unchanged.

In the practice of the IT industry, the objects are sets of large, interacting programs. In software, the determination of entropy corresponds to the task of measuring and, preferably, predicting the complexity of programs before they are developed. "Atomistically", the program is built of nested procedures and objects that call each other. The complexity of the product is determined by the number of possibilities offered by the programming languages that the programmer exploits or must exploit to accomplish his task. The goal is to work with human programmers to find structures that solve the given problem (i.e., meet the specification) and provide as little opportunity for human error as possible. A measure of software complexity must take into account this multitude of programming decisions— corresponding to the selection of characters from the character set in the case of text, or the location and velocity in the given space in the case of a particle in physics. Probably the simplest command for branching in a program is the GOTO command in many programming languages, which simply causes a jump in the program. It is the subject of an amusing historical dispute:

"'GOTO is considered harmful' is considered harmful."
Frank Rubin, 1987, after Edsger Dijkstra, 1968.

Branches, which the programmer puts into the program code, are inevitable core points for building complexity, but at the same time the sources of errors. They are a core part of the programmer's work. For natural philosophical purposes, the idea of so-called function points from software engineering is well suited. The technique was introduced at IBM in the 1970s as a way of measuring what is the real work in developing a program: The new. "Function points" are units in the program code, each of which adds a significant new functionality and increases the overall complexity.

In evolution, it is correspondingly the successful incorporation of a new ability or trait into a species. In large software systems, the work of thousands of programmers is coordinated at the points of function to form a functional whole.

The laws are similar to biological evolution, which constantly brings new functionality to organisms and increases the overall complexity of the biosphere, at least as long as biodiversity increases. This similarity is not accidental, evolution is a great collective software development.

In software practice, there is another significant effect. In sufficiently large systems, there are programming errors that need to be corrected (and many that are never noticed). The corrections are dangerous, the corrector often causes new errors. This also generally increases the entropy of the software system compared to the original system until the system becomes de facto unmanageable. Then you start again in software development with a new design. Let's hope that natural evolution does not do this resetting with us as mankind!

Thus, in three domains, we have defined a similar term that measures order (or disorder). It is in each case a kind of probability W of the possible states as a basis:

- Physics: W is the probability of all comparable states in space and velocity, the so-called phase space,
- Information: W measures the number of all comparable states with the characters of the information,
- Software: W is the number of states given by the number of inner program branches.

The current measure of entropy is then essentially given by the logarithm of this number. The clarity of the statements is weakest for software in view of the problems of measuring the complexity of software—nevertheless it has the greatest significance for life. Life is the complex of all processes taking place.

2.2.7 Entropy and Time

"The increase of disorder or entropy is what distinguishes the past from the future; this gives the direction of time."
Stephen Hawking, British physicist, 1968.

The Arrow of Time

Entropy reveals another typical property of randomness: Both are one-sided in time, the direction is clearly into the future. The time direction of the single coincidence gives the whole direction of the entropy development.

An illustrative term for this comes from the British astrophysicist Arthur Eddington (1928):

"The introduction of chance is the one thing that cannot be taken back. I use the expression 'arrow of time' to denote this one-sidedness of time. There's nothing like it with space."

The direction of chance gives the arrow of time: The arrow of time is one of the great statements and questions of physics. In contrast, simple, microscopic processes are reversible. This is clearly shown by the elastic impact of two balls: After the impact, the directions of the velocities could be reversed, the balls would run back, execute the impact inversely and return to the starting position. You can't tell from a film scene of a billiard shot whether the film is running in the right direction or backwards.

However, for some processes between elementary particles, to reverse in space, one must also replace the particle with the corresponding particle in antimatter to reverse the process as a whole in time.

The irreversibility of the arrow of time is particularly clear in the case of random machines. Figure 2.12 shows a sophisticated "random machine" for drawing the lottery numbers: Six numbers and an additional number are each released from the large container ball after quite a few revolutions, as randomly as possible. In all experience, it is absurd to

Fig. 2.12 The Swiss lottery drawing machine in action. (Image: Swisslos/own)

fill the balls back in, let the drum run backwards and expect the balls to eventually return exactly to their original order.

However, the whole world, indeed the whole universe, is a random machine and coincidences are part of nature. That is why we instinctively know the direction of the arrow of time when we see a scene of nature. For proof, let us think of a video showing the scene of Fig. 2.13: Colored drops of liquid, such as blood, fall one after another into water. Each drop immediately produces highly irregular colored streaks in the water. On the surface, circular waves probably run outwards. At the bottom one sees the transition to the uniform distribution of the red dye. The process starts from separated substances (blood, water) via chaotic phase back to order. The direction of the process is clear. If one would let this video run backwards, then we know it immediately: This is not the reality.

In the image we see irregular, random-looking clouds of color and on the surface the rings of the spreading waves.

In the macroscopic world of our everyday life there is only one direction: Entropy increases. A reduction is only possible artificially and with added energy. To remove the dispersed dye in the water, one needs a device (such as a filter membrane) and one must expend energy: The dye molecules are randomly mixed among the water molecules, there is a tremendous amount of randomness at any moment. Only identical ideal gases could in principle be unmixed without energy input and mix without energy release. This is an amazing little piece of physics about microscopic randomness, called Gibbs' paradox after the discoverer Josiah Gibbs:

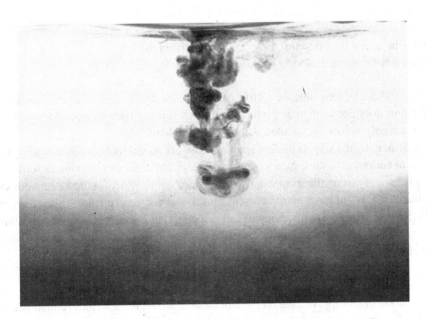

Fig. 2.13 Drops of blood falling into water. (Image: StockImages_AT)

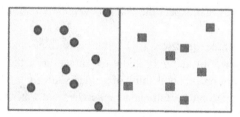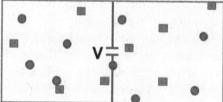

Fig. 2.14 On the mixing and segregation of two different gases and the work of the fictitious "Maxwell demon". Let "V" be a controllable valve. (Picture: own)

If two volumes of different ideal gases are mixed (Fig. 2.14), no energy is released, but the entropy increases by a characteristic measurable entropy of mixing. If, on the other hand, two volumes of the same gas are combined, the entropy does not change!

If one were to make the two different types of atoms to be mixed more and more similar in the classical manner in the thought experiment, the result would be a clash: the mixing entropy always remains the same, there is no other limit value. But suddenly, when they can no longer be distinguished, there is no mixing entropy. This incomprehensibly means a jump from finite mixing entropy to zero—classically, one physically expects continuity and continuity.

A fictional explanation, amusing but unphysical, is the segregation "by hand", by a demon. This demon, called "Maxwell's demon " after its inventor James Maxwell, sits at the valve in the center wall and exerts a control and works against chance.

To separate the mixture, the demon opens the valve

- When he sees a blue (square) molecule coming from the left, or
- When a red circular molecule approaches from the right.

After sufficiently long time the picture of the left side results again. But the unmixing of equal atoms does not work even in the thought-experiment. They are not distinguishable! Quantum theory solves the paradox without any problems.

The demon, by the way, is useless anyway: his work would make it possible to perform all sorts of curious activities, such as separating salt from water in a solution or sorting out hot particles and raising the temperature on one side—all of which would be economically valuable. But it would lower entropy just like that—and you can't do that. This is forbidden by the Second Law of Thermodynamics. But the mixing experiment reveals an intimate relationship between physics and information, hidden in the distinctness (or not distinctness) of particles.

The Heat Death of the Universe

The reduction of entropy, i.e., the creation of more order, requires an intervention from the outside. For systems on earth, also for us humans, this is possible and takes place continuously, e.g., in the living organism. Living organisms physically move outside the thermodynamic equilibrium with their environment, free energy[4] must be supplied from outside the organism. The physical processes increase the entropy. Our intellect can explicitly carry out or have carried out processes that lower entropy—e.g., tidying up a child's room. Only at death does the decaying body enter into equilibrium with the environment. When the atoms of the body disperse in the world, any old order is lost and the share of this former person in the entropy of the world becomes maximal.

Problematic (and rather philosophical) becomes the question of the fate of the whole world, the universe. In terms of entropy, the universe is (by definition?) closed and began in the Big Bang with the entire cosmic contents in an unimaginably densely packed primordial soup, enclosed in a volume the size of a football—that's how far back the past is visible to science (Siegel 2017). In our sense, this was an incredibly dense concentrate of randomness. Now entropy is continually increasing. Our biosphere is a temporary exception. In perhaps 100 billion years, the stars will have burned up and more and more black holes will have formed. The universe will be full of disorder and with no more free energy for life.

This end-time idea is older than the concept of entropy itself. The first idea came from the Irish-Scottish physicist William Thomson, called Lord Kelvin (1824–1907). Kelvin was one of the first to determine the temperature of absolute zero, introducing a temperature on a thermodynamic basis without reference to a specific material such as water. The unit of measure of temperature, the kelvin, is named after him. He had also thought about the age of the sun (and earth). Specifically, he had reduced the age of the sun in several estimates to finally 20 million years. This numerical value was almost a disaster for the development of science, because Lord Kelvin opposed Darwin. 20 million years do not fit to the development of the species. But it was the first time that physics in this fundamental sense was ever applied to the sun and the earth.

20 million years is quite different from the true value, about 4.6 billion years, but he did not know the energy source of the sun! For the universe as a whole he foresaw a slow death by cooling: All mechanical energy becomes heat, nothing happens any more, especially no new stars are formed. The catchword of heat death is to be understood in this sense, not as a high temperature.

With the concept of entropy (and thus of total disorder), the idea of heat death received a more scientific basis. Since, according to this thought, the temperature of the universe tends to cool down, perhaps approaching absolute zero, one could probably rather speak of cold death. Especially if the expansion of the universe continues or even accelerates. To the beginning of the universe, the Big Bang or Big Freeze, then the "Big Chill" or "Big Freeze" occurs as a vague end in distant times, in 100 billion years or more. Then chance would be

---

[4]Free energy is the fraction of total energy that can do work. It decreases when entropy increases.

frozen in the cosmos. But here science becomes postmodern physics and speculation as we defined it above, or natural philosophy.

2.2.8   Summary of the Chapter

Even the word "coincidence" or "zufall" is problematic. In many languages it is emotionally loaded, indeed there is a goddess of fate behind it. We define coincidence neutrally, in the simplest case as a causal chain which is one-sided and starts at a certain time. There is no information about the past before that. On the other hand, we tend to think in terms of continuous processes. Then not only are there no jumps, but the trajectories become as smooth as possible. However, even with smooth curves, chance enters the world. On the one hand, because of the initial conditions, which can never be exactly fixed and repeated, and on the other hand, because of the many players, all of which interact and are also subject to chance. This is especially true of the important example of our solar system. Such systems with apparent randomness despite determinism are called chaotic: what the observer then sees and cannot understand, he thinks is random. The solar system, our emotional reference for stability, is also chaotic.

Chaotic systems are an example of how a tiny coincidence can have an arbitrarily large causal effect; the much-strained image of the butterfly in Brazil belongs here. Using Brownian motion, we construct a simple apparatus that converts an atomic motion into an arbitrarily large event. This is proof that a single quantum fluctuation can have a noticeable effect in the macroscopic world. An accident in the world of atoms can unhinge the big world.

Invisible to the eye, our macroscopic world contains vast amounts of randomness. The macroscopic measure of this is entropy.

Entropy is a rather abstract quantity in thermodynamics; in the microscopic world it becomes plausible as a measure of the degree of disorder of a set of objects, or of the information necessary to describe them. As a type of entropy, we consider the complexity of a software system. Here it is the set of internal decisions in the program that constitute entropy.

Chance and entropy define the direction of time. We humans have learned this in everyday life (or evolution). Mentally, a mind could try to reverse the effect of the direction of time, but it doesn't work. Still, the idea of this Maxwellian demon is a neat meme in physics. The demon shows us the connection between physics and information.

Time also runs on for the universe as a whole. Entropy continues to grow and the result is (probably) a standstill of all activities. The catchword of 'heat-death of the universe' is also a physical meme since the nineteenth century. One expects today by the expansion of the universe thereby rather a cold death, a Deep Freeze: The maximum of coincidence is frozen. But these are postmodern speculations.

References

Chaitin, Gregor. 1969. On the length of programs for computing finite binary sequences: statistical considerations. *J ACM*. 16: 145. https://doi.org/10.1145/321495.321506.

Eddington, Arthur. 1928. *The nature of the physical world*. London: Macmillan.

Eigen, Manfred, and Ruthild Winkler. 1975. *Das Spiel. Naturgesetze steuern den Zufall*. München/Zürich: Piper.

Faye, Hervé. 1884. *Sur l'origine du monde*. Paris: Gauthiers-Villars.

Kaiser, Peter. (1990). Die Lösung des Einstein Kausalitätsproblems. www.max-stirner-archiv-leipzig.de/dokumente/Kaiser-Einstein.pdf. Zugegriffen im Mai 2020.

Kolmogorow, Andrej. 1965. Three approaches to the quantitative definition of information. *Problems of Information Transmission* 1: 1–7.

Monod, Jacques. 1970. *Le hazard et la nécessité. Essai sur la philosophie naturelle de la biologie moderne*. Paris: Le sueil.

Siegel, Ethan. (2017). *How big was the universe in the moment of its creation?* Forbes.com/sites/startswithabang. Zugegriffen am 24.03.2017.

Sussman, Gerald, and Jack Wisdom. 1992. Chaotic evolution of the solar system. *Science* 257: 56–62.

The Natural Coincidence

Q: **Is the slight noise floor on the active speakers normal?**
A: **Yes. They are allowed to murmur very quietly.**
Musiker Board Forum, 2019.

Everything rushes, everything is noisy. Elementary noise has become audible with electrical engineering and electronic amplifiers. In this sense, the term has existed since the work of the German physicist Walter Schottky in 1918. We extend the notion from (electro-) acoustics and the noise of an amplifier in general to a restless but constant stochastic[1] disturbance in the background. The first mathematician to study disordered, curly nature with randomness was Benoît Mandelbrot.

3.1 Looking Correctly According to Mandelbrot

"**Clouds are not spheres, mountains are not cones, coasts are not circles, and bark is not perfectly smooth, nor does lightning travel in a straight line.**"
Benoît Mandelbrot, French mathematician, 1924–2010.

The meaning of Benoît Mandelbrot's saying is the invitation to take a closer look at the world. It is much more finely constructed than we normally imagine. The quote warns against the usual reduction of real, complex objects to simple bodies. But precisely this abstraction was a condition of the success of Western science in the style and sense of Aristotle. The process of abstraction is humorously described by the physicist Arthur Eddington using the example of a physics examination task with the "Eddington elephant":

---

[1] Stochastic means "influenced or determined by chance". See glossary.

© Springer Fachmedien Wiesbaden GmbH, part of Springer Nature 2021
W. Hehl, *Chance in Physics, Computer Science and Philosophy*, Die blaue Stunde der Informatik, https://doi.org/10.1007/978-3-658-35112-0_3

"An elephant slides down a grassy slope . . ."

and the student translates

"A mass point slides down an inclined plane at an angle of 30° with a coefficient of friction μ = 0.05."

Eddington comments laconically:

"All the poetry of the task is gone".

Much of the poetry of the task is accidental and unimportant to the solution. The Romantic and Platonist Johann Wolfgang von Goethe would certainly agree with him, but one can reckon with abstraction and begin to understand. Understanding nature by abstraction to mathematics, more precisely by abstraction to geometry, was expressed by Galileo in 1623 in one of his best and most famous quotes (Hehl 2018):

"[The book of nature, philosophy] is written in the language of mathematics, and its letters are circles, triangles, and other geometrical figures, without which it is impossible for man to understand a single image of it; without these one wanders about in a dark labyrinth."

The origin of the idea of the quotation is (as Galileo himself suggests) the alleged saying over the gate of Plato's Academy:

"Let no one enter my door who does not know geometry",

related by the quotations (after Plutarch), *"God is the great geometer"* and *"God forever drives geometry"*- *(Ἀεί θεός γεωμετρε⬚).*

But both the modern physicist Eddington and Galileo in the late Renaissance are aware of the limits of abstraction. Arthur Eddington compares the physicist's research somewhat extremely to the actions of the legendary giant Procrustes in Greek mythology. Procrustes offered travelers a bed. If they were too big, he chopped off their feet; if they were too small, he stretched their limbs. Eddington adds, *"after this, he wrote a scientific treatise 'On the Constant Length of Travellers'".*

Mandelbrot means in his quote above that when we look at the world, we mainly see regular structures that we know and that we simplify what is actually complex. We pay little or no attention to the subtleties of chance. Euclid's basic geometric structures are smooth and simple.

Mount Schiehallion in Scotland (Fig. 3.1) entered the history of science because it is relatively close to a regular cone shape. It was used in 1774 to measure the deviation of the plumb line through the mountain and thus to determine the gravitational constant and mass of the Earth. It was chosen as the "smoothest" mountain in Great Britain for the experiment.

Fig. 3.1 The Schiehallion. A mountain is sought as a smooth geometric cone. (Image: Schiehallion01, Wikimedia Commons, Andrew2606)

Higher mountains deviate more from simple geometric shapes, such as the ridge of Fig. 3.2. The contours are visibly full of randomness. One sees more and more "coincidence" the closer one looks at the terrain. Of course, the Schiehallion also shows irregularities in the contours on closer inspection. The mountain is the site of a second scientific pioneering act: Schiehallion was the first terrain in history to be surveyed and mapped with contour lines.

When "looking more closely", problems arise with natural objects (and with specially constructed mathematical ones). This has become known historically as the "paradox of the length of coasts". For the first time, the British meteorologist and peace researcher Lewis Richardson reported it.

Richardson was quite an extraordinary person and a pioneer in the calculation of stochastic processes, or rather he identified various processes as random but still calculable. These included the weather, but also wars and the observation of the "real" lengths of real curves such as coasts or national borders. He was the first to mathematically analyse wars between nations, but also criminal gangs, and to model them as coincidences (as a Poisson process) and to find a distribution law for the size of wars (number of dead) depending on their frequency. Richardson is thus considered the inventor of peace research.

More concrete was his pioneering work in mathematical weather calculation. To this end, he proposed in 1917 to solve the fundamental equations of fluid mechanics and

Fig. 3.2 Typical rough high mountains: The Mürtschenstock, Switzerland. (Image: Edith Geissmann)

thermodynamics numerically, long before the availability of computers! He thought of "original computers", i.e., human computers:

> **"Richardson took six weeks to make a forecast for six hours ahead - and it was still wrong. He explained that 60,000 human computers with slide rules would be able to calculate the weather just as fast as it arrived."**
> **Lewis Richardson according to BBC News, South Scotland, 2013.**

Let's assume that Richardson used this to calculate about 10,000 operations per second. Today's computers with weather software usually calculate the deterministic randomness of the weather sufficiently for five days ahead. However, the computing power of weather computers is one million times one million that of the above fictitious team, and this without error and highly accurate.

Another phenomenon with coincidence, which at first looks firmly determined, is the length of natural lines. Richardson had noticed that the length of the border of Portugal with Spain was given as 1214 km in Portugal, but only 987 km in Spain. He concluded: Lines with coincidence (from a mathematical point of view, not historical) are indeterminate. The result of the measurement depends on the observer.

The most extreme case of no coincidence is the simple straight line so demonized by the painter Friedensreich Hundertwasser

"The straight line is ungodly and immoral. The straight line is not a creative line, but a reproductive line. In it dwells not so much God and the human spirit as the convenience-loving, brainless mass ant."
Mold Manifesto, 1958.

The path of light in a vacuum follows a straight line, indeed it defines the straight line for physics, even if the light beam bends because of a gravitational field. But material "straight lines" in nature are more complex: the lines of the hand, for example, the trunk of a tree, the path of a lightning bolt, or even the line of coast. Only very carefully drawn crystals show straight edges and form ideal bodies—usually, natural crystals are also full of coincidence in the form of various kinds of errors such as dislocations, of atoms that do not belong to the crystal and of empty spaces where atoms should actually belong. The linear structures of nature are more or less curly, the surfaces rough, the contours jagged in space.

Figure 3.3 explains Richardson's paradox of coastline lengths using the example of the British Isles. The smaller the applied yardstick, the more curved the contour becomes and thus the longer the measured coastline. Finally, if the scale is too small, the measurement

Length 2800 km   Length 3400 km

Fig. 3.3 Illustration of the paradox of the length of coastlines. The examples illustrate the "length of the coast of Britain" measured with 100 km and 50 km scales. (Image: Coastline 100 km and 50 km, Wikimedia Commons, Avsa)

becomes meaningless, for example, if the wave action on the beaches does not allow a more accurate measurement. The randomness of the structure again means uncertainty. The degree of increase in length with the refinement of the unit of measurement characterizes the roughness or smoothness of a coastline.

The French-American mathematician Benoît Mandelbrot used Richardson's coastal example to establish a field of mathematics that deals with non-smooth structures in nature, in economics, and especially in mathematics. In 1975, he coined the word fractal for this from the Latin *fractus* 'broken'. We will explain why this word in a moment. It is one of the most successful popular terms in science ever. Fractals exist in mathematics and in nature, as we will see in a moment, but also in finance in stock prices and in the humanities. Even the Bible has been taken to be a fractal, and reading the Bible has been seen as a fractal theological experience (Brookman 2015), and the alternation of peacetime and wartime (Braden 2009) has been modeled in esoteric fractals.

In mathematics with the idealized structures, one speaks of fractals, in nature better of "fractalesque" structures or random fractals, because natural objects also contain intervening chance. For the systematic construction of fractals one uses in principle again and again the same structures in changed scales; thus a certain or in mathematics also strict self-similarity arises. Fractals have become very popular, also because of the beautiful pictures. The already mentioned John Archibald Wheeler said in 1982:

"No one will be able to call themselves scientifically literate who doesn't know about fractals."

Figure 3.4 shows one of the most beautiful examples from nature, the flowers or florets of broccoli romanesco. They are wonderful mathematics with approximate self-similarity

Fig. 3.4 Natural fractal Brassica romanesco. (Image: Wikimedia Commons, Richard Bartz)

over about three stages, cone on cone on cone, and in addition almost perfectly implemented Fibonacci numbers in the numbers of spirals emanating from the top of the cones. It is wonderful mathematical regularity with randomness. We recommend the reader to look at the picture in high resolution!

One of the simplest and best-known fractals in mathematics is the snowflake and curve devised by the Swedish mathematician Helge von Koch in 1904, the "monster curve" (Fig. 3.5). You can see the construction principle: Every straight line piece is replaced by a jagged piece of the same length with an inserted equilateral triangle, over and over again. Only the kink points remain ad infinitum.

The generalization is instructive, if the inserted piece is not inserted with 60°-angle, but in another angle. These so-called Cesàro curves are illustrated in Fig. 3.6. The length of the curve grows to infinity, while the enclosed area has a finite limit, namely 9/5 of the original triangle.

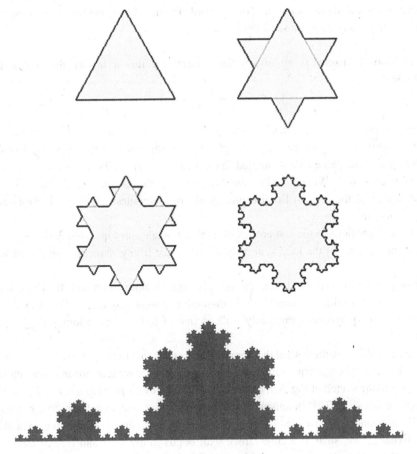

Fig. 3.5 Artificial perfect fractal Koch curve. (Image: Wikimedia Commons, Koch flake and Koch curve)

Fig. 3.6 The Cesàro curves for angles of attack from 0° to 90°in steps of 10°. Explaining the paradox of the length of coastlines. (Image: Wikimedia Commons, Fractalgeometry123)

The transition from the straight shape at the top left corresponding to 0° to the filled triangle at 90° shows how the curve increasingly fills the area. This effectively means the transition of the dimension of the curve from 1 (normal line) to 2 (area) and meaningfully with fractions in between. It is the fractal dimension or Hausdorff dimension after the German mathematician Felix Hausdorff as a measure of the deviation of geometry from smoothness. It is also the background of the word "fractal". Using the Hausdorff dimension as an effective dimension, one can define.

▶ **Definition A fractal is an object that effectively has a higher dimension than topological.**

Thus, natural objects also receive ratings: The more randomness or disorder, the more the effective dimension increases. A straight line has dimension 1.0, angled lines more, a line finally fills the plane with its angled lines, then it effectively becomes a surface with a dimension close to 2. We are talking about natural objects that are "curly", "frayed", "full of bubbles", and the like. In English one speaks of "wigginess". Table 3.1 shows some values in nature.

Some explanations to the last entries of natural fractalesques in Table 3.1.

The internal area of the lung is already almost space-filling: the effective dimension is almost 3.

Brownian motion is equivalent to "wandering" (much more common is the English term random walk) in a grid, for example of north-south and west-east streets (Fig. 3.7). At each intersection, the person can randomly walk in one of the four directions, i.e., also back again.

The astonishing mathematical theorem of Pólya states that in this odyssey every starting point will be reached again "sometime", one only has to wander around long enough. Therefore, in the sketch of Fig. 3.7, one can also think of every point as a starting point—all points are equivalent. This statement of safe return is, of course, even more true in the case of a linear wandering back and forth on a straight line, but it is not true in a three-dimensional lattice, such as a cube lattice with bonds in the x, y, and z directions; there the wanderer returns by chance only 34% of the time. The reason is that there are too many paths in space that "one" (i.e., chance) can choose.

Table 3.1 Some Hausdorff dimensions of fractal or fractalesque objects in nature, i.e., of natural structures with a high content of disorder

Length of the coast of Britain	1.25
Australia	1.13
Norway	1.52
Ireland, west coast	1.22
Ireland, East Coast	1.22
Ball made of crumpled paper	2.5
Broccoli, surface	2.7
Cauliflower, surface	2.3
Brain, surface	2.7
Lungs, internal surface	2.97
Brownian motion or random walk In two and more dimensions	2.0
Snow needle shaped crystals	2.1
Star-shaped crystals	2.4
Sleet	2.9
Lichtenberg figs. (3-dimensional)	2.5

Source: Wikipedia article "List of Fractals by Hausdorff dimensions"

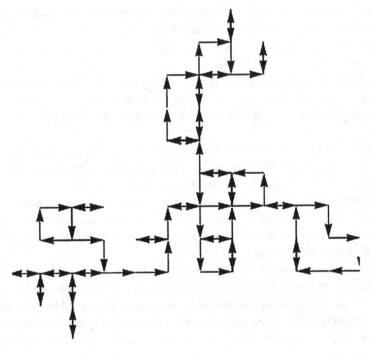

Fig. 3.7 Extract from a 2-dimensional odyssey in a regular road network. (Image: Fabio Vanni, Sciencespo)

In many situations, chance chooses its direction of action not neutrally, but with preferred directions, as in rolling a "loaded" die that does not produce the ideal equal distribution of the rolled eyes one through six. We call this directed randomness. In the section Evolution we discuss the philosophical meaning in a little more detail.

A particularly beautiful example of such directed coincidence in nature is the formation of snowflakes and snow formations. The process of formation is complex: Above a temperature of $-35°$, snowflakes need a nucleus to form. Tiny water droplets attach themselves to crystallization nuclei as ice and grow by further attachment of water molecules to the outer edges as they fall rotating through the cold, moist air. The smallest vortices are created by the released heat and generate the rotation. The preferred directions of growth are dictated by the directions of the hydrogen bonds; thus angles of 60° or 120° are produced (see Chap. 6). This symmetry creates a strong impression of self-similarity. A snowflake weighs about one milligram and consists of about 10^{19} water molecules. Thus, de facto, the random product "snowflake" never exists twice in identical form. The beautiful highly symmetrical photos of snowflakes are deceiving (Fig. 3.8); most snowflakes are only partially regular. Coarse randomness interferes with directional symmetric randomness. Depending on the fine structure, the flakes then clump together in different densities as indicated in the Table. A snowdrift is a gigantic random machinery visible to the naked eye or at least with the light microscope.

The final example in Table 3.1 of objects with a high degree of randomness are the *"aesthetically pleasing tree-, fern-, or star-shaped patterns that occur on insulating materials during high-voltage electrical discharges"* (German Wikipedia). These fractal-like structures were discovered by the witty (and quote-rich) German physicist Georg Christoph Lichtenberg around 1778. When a high-voltage discharge occurs, including lightning, the forced currents cause fractures in the material, or burns, which persist and cause real works of art (Fig. 3.9). However, also on the human body in people struck by lightning!

To complete the picture, here is a living electrical random experiment, the so-called Tesla sphere, invented in 1892 by the Serbian-American inventor Nicola Tesla (Fig. 3.10). In the centre of the gas-filled sphere is an electrode that sends high-frequency alternating currents into the sphere. This triggers several current filaments, luminous channels from the inner sphere to the outer. When approached with the hand or finger, they react sensitively. The lamp exerts a great fascination through the liveliness of the proliferating filaments and the dynamic guided coincidence.

The mathematics of fractals allows to investigate the geometry of chance with mathematical methods. In nature, there is randomness everywhere; the table shows only a few examples. Figure 3.11 illustrates the difference between mathematical fractals, which nest structures indefinitely deep, and physical reality, with a limited range for similar structures. Finally, in nature, with sufficient depth, atoms become visible and end the freedom to build arbitrary structures.

The view starts at the top with the normal view without magnification. The arrows pointing downwards indicate that the scale is getting smaller as you look at it. Step by step,

Fig. 3.8 Shapes of snowflakes. From the "Book of Nature" by Israel Perkins Warren, 1863. (Image: Wikimedia Commons, ComputerHotline)

therefore, is enlarged by one order of magnitude. "At the bottom" (after an expression by physicist Richard Feynman) are the atoms and molecules with their interactions. If one goes "down" to the size of atoms, this corresponds to magnifications of up to ten million times in order to clearly see hydrogen atoms, for example.

The left side of the graph demonstrates the self-similarity of mathematical fractals, which repeat their structure indefinitely. For the mathematician, each procedure for a new type of fractal is an adventure in itself. In nature, self-similarity occurs in a wide variety of

Fig. 3.9 Lichtenberg figures created in an acrylic sculpture. (Image: flickr/Wikimedia Commons, Jeff Keyzer in Ada's Technical Books)

physical environments. However only over finite ranges, because the acting forces are scale-dependent and especially the relation of the acting forces among each other changes with other sizes.

Galileo Galilei was probably the first to make this fundamental observation in 1632 when he compared the construction of large ships and boats in the arsenals of Venice: For the construction of larger ships, he observed disproportionately large, stable scaffolds to let them into the water, which are not needed for small ships. Accordingly, he sees in nature that bones of large animals have different proportions than bones of small animals (Fig. 3.12).

To this end, Galileo discovers a puzzle that corresponds exactly to our diagram in Fig. 3.11, the difference between objects of mathematics versus objects of physics. Circles always remain circles, no matter whether they are large or small? However, as Galileo points out, in physics it is the scale that matters:

> **"But because all mechanical causes have their foundation in geometry, where the size or smallness of circles, triangles, cylinders, conoids, and certain other formations do not matter ..."**
> **Galileo Galilei, Discorsi, 1638.**

Galileo wonders why a horse cannot jump as high as a flea in relation to body sizes? We know that a fundamental reason for the different size scales is the existence of atoms (Galileo believes in atoms, but he does not mention them in this context).

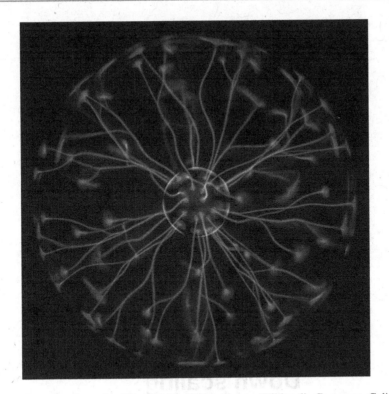

Fig. 3.10 A plasma sphere after Nikola Tesla (1892). (Image: Wikimedia Commons, Colin)

If on the scale while "going down" to smaller and smaller the atoms become visible, the laws change from the continuum to atomic laws and thus the manifestations of all effects. This is the solution to Galileo's riddle. This thus also applies to the occurrence and nature of chance in the various magnitudes of things down to quantum effects: The laws of chance also change.

There is much more coincidence in the world than the examples given so far.

So far, the very special and most famous "fractal" is missing: The Mandelbrot set (Fig. 3.13). The main reason is that the Mandelbrot set contains no randomness, but is strictly determined. Two equations (one equation when using complex numbers) are applied over and over again: If the result for a point explodes, the point does not belong to the set. The resulting image, popularly known as the apple man, is a two-dimensional world with infinite subtleties, determinate yet arbitrary-fantasy looking. Mandelbrot shows how a simple rule can give rise to a complex-looking world.

We have seen two criteria for fractals: Typical patterns are systematically repeated in a fractal (self-similarity) and the effective dimension of the object (the Hausdorff dimension) is larger than it should be in normal observation (the topological dimension).

Self-similarity is present in the image in a complicated sense: Even if one reduces by a factor of 10^{100}: 1, the same types of structures always appear like "leitmotifs". They are, for

Normal view

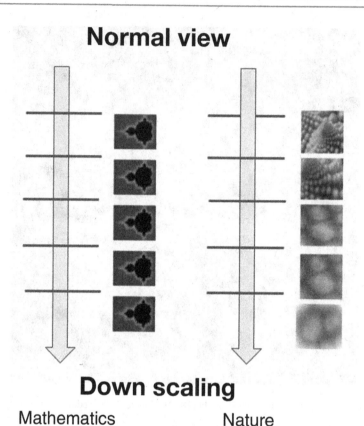

Down scaling

Mathematics   Nature

Fig. 3.11 Illustration of the difference between "real" fractals in mathematics compared to "fractal sinks" in nature. Each step down symbolizes an order of magnification. (Image: own)

example, typical snails (Fig. 3.13b) or the basic structure itself, the apple manikin. However, except for a few places, they are not exactly identical.

The dimension of the object Mandelbrot set is on the one hand the boundary curve of the set itself, on the other hand the area. The boundary curve, as a curve actually of dimension 1, has by its complex structure the Hausdorff dimension 2 and is thus fractal. The enclosed area has both topologically and according to Hausdorff the dimension 2 and is therefore according to Mandelbrot, because the two values are equal, not fractal. The length of the boundary curve is unbounded ("infinite"), the enclosed area is finite, approximately 1.506484, and could not yet be determined exactly.

The Mandelbrot set is the most complicated object we know: A virtual flight into its depths, a "deep zoom" into this pseudo-random world, with the help of a high-definition video is an intellectual, almost spiritual pleasure.

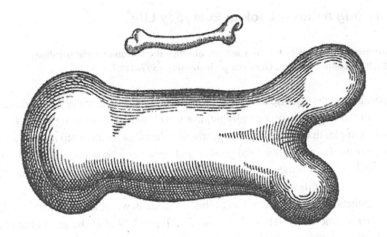

Fig. 3.12 Limited self-similarity in reality. Galileo Galilei, 1632 (Image: Dialogue on Two New Sciences, Fig. 27. Online Library of Liberty)

General view of "The Little Apple Man
Image: Blue-Gold Mandelbrot set,
Wikimedia Commons, ArEb.

Very enlarged section
Image: own and Peitgen, 1986.

Fig. 3.13 Two views of the Mandelbrot set. The color and the black and white codings indicate the local velocity at which the calculus tends to infinity at a point on the number plane

3.2   Taking a Closer Look in Everyday Life

"Fortunately our eye is not a microscope, already the common sight idealizes"
Friedrich Theodor Vischer, German philosopher, 1807–1887.

The classical aesthete Friedrich Vischer (Sturm 2003) thinks that only the whole can be aesthetic and that the "microscopically adjusted eye" only discovers ugliness. By this he means, for example, that a woman's body, whether "real" or formed into a statue in marble, can only be beautiful as a whole and perfect, but seen microscopically with skin pores or marble hatchings it becomes ugly. Naturally, even smooth skin becomes a furrowed landscape when magnified, but microscopic or even electron microscopic images have their own aesthetic appeal. The sources of this are often new, unfamiliar shapes and, above all, the immense amount of detail "down there". Figure 3.14 also shows marble in greater detail. It is a thin section in a light microscope with polarized light, which is responsible for the magnificent colors. What can be seen is colorful randomness in the form of crystals in random shapes and orientations.

Many images in the electron or scanning electron microscope are of fascinating beauty. But the philosopher is right: We are used to idealizing the objects we see and to

Fig. 3.14 A thin section of calcite crystals in polarized light at low magnification. (Image: Bernhard Lebeda, in mikroskopie-forum.de. With friendly permission)

overlooking deviations from the ideal as long as we can. In this sense, we are Platonists like the aesthetic philosopher Vischer quoted above.

3.2.1 The Normal Coincidence Around Us

"Tiny things ("les petits riens") are never entirely without meaning; beauty abounds in the tiniest."
Sylvie Germain, French author, born 1954.

The little coincidence is all around us, mostly unnoticed and without effect. Sometimes it is even beautiful or adds to the beauty or sense of the big. Sometimes we see it with the naked eye, sometimes, such as when we slide our finger on a surface, we feel it (at least as long as the roughness of the surface with its mountains and valleys is not less than 1 μm). But it is always there. That should be disconcerting to the rationalist.

It was a private communication by Benoît Mandelbrot: He was concerned with the fact that we are surrounded by chance or *Zufall* everywhere. Mathematicians and physicists had ignored characteristic traits because there was no mathematics for them before, namely no theory of fractals yet. Mandelbrot saw the world full of fractals, from the galaxies to the water levels of the Nile and the prices on the stock exchange.

Figure 3.15 shows a scene from everyday life with objects from nature and artificial things, all with fractal details:

- Plants, such as roses, trees, grass,
- sky with clouds,
- a lake,
- artificial objects like houses and tables.

Nanostructures and Nano-Nature
All these objects are indeed full of chance: The exact structures of the plants, the clouds and the waves on the lake. But also the man-made objects, you just have to look closer, take a magnifying glass, a light microscope or even a scanning electron microscope. Here is an illustrative comparison (Fig. 3.16).

It is an image of a scanning electron microscope in which a fine metal tip electrically scans a surface. The image shows an institute logo written in molecules of a plastic on a graphite surface. The width of the grooves is about 1 to 3 nanometers. The plastic molecules are clearly visible, whereas the carbon atoms of the graphite are only 0.3 nm apart.

The effective magnification of the image is about 1: 1,000,000! At this magnification the world looks different; here are some examples:

Fig. 3.15 An everyday scene with different types of objects. All objects carry coincidence when looked at closely

Fig. 3.16 A logo written in molecules. Scanning tunneling microscope image of molecules on graphite. (Image: Wikimedia Commons, Frank Trixler, LMU/CeNS)

- My laptop will be as big as Switzerland.
- One key of the keyboard becomes a square with 10 km edge length,
- A hair becomes a colossus with a diameter of 50 m.
- The grooves (10 μm) that can still be felt are ten meters deep.
- The spheres of the corona virus (between 50 and 200 nm) become tennis balls and footballs.

At this magnification, the atoms are still small, about the size of pinheads! In our magnification comparison, a leaf from an oak tree would be perhaps 50 km long. Even with the naked eye, the philosopher Gottfried Leibniz could not find two identical leaves on a tree in the garden of Herrenhausen Palace in 1675: How much more possibilities for differences does the "leaf landscape" offer when viewed at the level of atoms! The construction of the macroscopic world from the atoms of this smallness makes possible an unbelievably great diversity in the world and—for example, in us humans—an incredible individuality.

Why has nature made us humans so huge, or had to make us so huge, compared to the basic atomic building blocks?

The reason lies not directly in physics, but probably in the necessities of information technology. Nature needs such a large infrastructure to build a computer with the power of the human brain, even in such compact biological technology with electrochemical neuron cells. After all, it takes nearly 90 billion neurons! Thus, we have two landmarks for the construction of intelligent life: the size of atoms and the size of the brain. Whereby *the* expression *"nature needs"* is a problematic human expression to describe the way evolution works.

Probably the first to become aware of the smallness of the atomic world and its possibilities (and its laws) was the American physicist Richard Feynman, in a famous speech in 1959 entitled

"There is plenty of room at the bottom"

But going down (to atomic dimensions) has consequences:

- The requirements for the accuracy of a device to be built become higher. In a larger device, the tolerance range for a machine to work may be many billions of atoms more or less—the machine will still work. In a nanomachine, all atoms may need to be in the right position, none too many or too few.
- The fluctuations become more frequent and relatively larger as the structures become smaller. The driving disturbances, such as the average thermal energy per degree of freedom, also become relatively larger.

Plants and Chance

The closer one looks, the more randomness becomes visible, for example in the house walls and the table surface of Fig. 3.15, but also in the structures of the leaves and the trees. One recognizes fractal structures. It is everywhere chance with a guiding principle, in the tree in the leaves and the branches.

The task that nature solves in the pseudofractals of Fig. 3.17 is an optimisation problem: the aim is often to provide the largest possible leaf area and to optimally spread and distribute juices. The constraints are the minimum possible consumption of material and the best possible mechanical stability of the whole. Under this mantle of an approximate optimum, the multitude of typical leaf shapes of thousands of species emerges, and under the mantle of one species, the multitude of stochastically different individuals. By analogy with physics, we call these smaller, multifaceted changes a noise; we will encounter this term many times.

We also call these small variations the small coincidence or *weak zufall,* which arises with the slightest fluctuations and mutations at the micro-and molecular level. This coincidence is built into all organisms, also into us, into our body, and therefore also into our soul.

In addition to Richard Feynman's saying, one could say with the view of chance:

"There is plenty of room at the bottom for randomness."

Thus, it also has room for individuality and progress (see evolution). This includes as a decisive, small coincidence for us, that exactly *the* sperm reached the available egg, which let us come into being. To the degree of the acting coincidence a quotation from the German Wikipedia article "Sperma" (pulled July 2020):

"On average, the volume of a human ejaculate is 2 to 6 ml, with 1 ml containing an average of 20 to 150 million sperm (cf. 200–300 million in the stallion)."

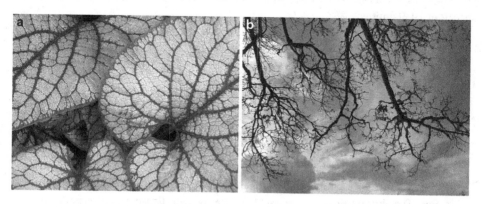

Fig. 3.17 (**a**) Leaf structures. (Image: Fractal Pattern Leaves, Wikimedia Commons, Laurenjessiehatch.) (**b**) Structures of stems and branches. (Image: Tree Fractal, Wikimedia Commons, Laureenjessiehatch)

It actually has a lot of room for happiness and unhappiness, for health and specialness, for intelligence and stupidity down there.

3.2.2 · Water Waves and Randomness

Interviewer: And why is the water moving? Child: I don't know.
Interviewer: OK. And how does it move? Child (yawns): There are just waves.
From: Interview study, Moritz Halder, 2017.

The explanation is not that simple—the concrete wave phenomenon on the surface of the water includes oscillating players or better counter-players and in addition a generator that provides the energy for the waves. Water waves can show us vividly how chance grows. We consider here only wind-generated water waves, i.e. water surfaces of rivers, of lakes or of the sea, which are deflected by the wind. There are two types of water waves: capillary waves and gravity waves (Fig. 3.18).

In the case of very small waves, it is the surface tension of the water that tries to pull back deviations from flatness. Surface tension turns the water surface into a taut cloth: an upward deflection is pulled down, while a downward deflection is pulled up. Capillary

Fig. 3.18 (a) Capillary waves: Cat Paw's in a Norwegian fjord. (Image: Wikimedia Commons, Blue Elf). (b) Gravity waves on the ocean: research vessel Delaware II in a storm. (Image: Wikimedia Commons, National Oceanic & Atmospheric Administration)

waves are waves with a wavelength of a few centimetres and a propagation speed of 20–30 centimetres per second. They are the ripples on the water surface when wind rises after calm, or the rings of waves when raindrops fall into the water. It takes a minimum wind speed of about 70 cm/sec to even start waves. This gives the shortest possible waves of about 2 cm wavelength. Capillary waves are often easily visible due to the light reflections or distorted reflections in the water. A maritime term for these gentle waves is cat's paws. The random character of the waves is gentle and peaceful.

In the larger waves, gravity is the decisive factor. In these waves, the energy oscillates between potential energy (the distance from the zero level of the water surface) and kinetic energy (the kinetic energy). The laws of these gravity waves are completely different from those of capillary waves. Thus the waves can become so steep that they become unstable and the wave "breaks", especially in shallow water where the waves are slowed down. But these waves also have laws that they follow, on the one hand, and a stochastic character from the moment they form. Thus it is a law that in deep water a group of running waves runs half as fast as the single wave: It is therefore looks as if a new wave emerges behind a group of waves, would grow, run through the group and disappear at the front of the front.[2]

Much about waves is random: From first beginning at calm (however unstable) border surface of water and moved air, from transfer of air vortices to capillary waves up to collapse of small waves and transfer of energy to large waves up to 15 metres height. For this to happen, the wind has to blow consistently and strongly from one direction (it's a kind of great coincidence), e.g., at about 90 km/h for three days, to reach the 15 metre wave height (from the bottom to the crest). The boiling sea becomes a dramatic performance of chance and our being at the mercy of it. For the sailor, even the waves on a lake in a storm are an impressive (and fear-inducing) demonstration of disorder.

The "most terrible coincidence on the ocean" is the observation of a regularity in coincidence made by the author himself:

"If waves have a certain extraordinary wave height in a storm, they will eventually reach a double wave height."

We had often taken walks along the coast of the sea ("only" the Mediterranean), especially during storms, and watched the waves coming up far out and the spray hitting the rocks. A wonderful, impressive spectacle. But after a storm, the first time the waves broke much further out, reaching places at the far edge of the sandy beach, and the spray was twice as high as ever!

This means that many statistical distributions of phenomena have a "long tail", i.e. there are still possible events far outside the normal. In the case of ocean waves, this means the existence of monster waves (or rogue waves).

---

[2] The group velocity is half the phase velocity.

One of the most famous images from Japan, "the great wave of Kangawa" (Fig. 3.19) is an artistic representation of this coincidence in such a threatening form for the fishermen's boat. Monster waves are rare, unpredictable, and of tremendous destructive force, many times the usual largest waves. Ships are built for waves that crash at up to 15 tons per m² of ship's surface. The usual largest waves break over the ship at 6 tons per m². Chance builds monster waves by "constructive interference" that come over the ship at 100 tons per m² and 70 km/h. The largest measured height of a monster wave is 30 metrès. There is no warning, possibly preceded by a phase of short calming. It is a threatening coincidence for all ships after several days of the storm.

Wikipedia writes in the article "Rogue Waves" (pulled July2020):

"One of the remarkable features of monster waves is that they always appear from nowhere and quickly disappear without leaving a trace."

At sea and with little wind, the wave process like poetry, on the ocean, it is a deadly game of chance. But even on the gentle lake, the waves are almost always a natural theatre and lesson randomness in their variety of structures that can be observed and that are constantly changing over the entire surface. There are even small relatives of the monster wave on lakes, called "solitons". These are single waves or short wave packets that run across the

Fig. 3.19 An artistic rogue wave. "The Great Wave of Kangawa" by Katsushika Hokusai, 1830 Tokyo National Museum. (Image: Wikimedia Commons, Google Cultural Institute)

lake as if from nowhere and do not decay for a long time. The physical explanation for monster waves and solitons is non-trivial (if "trivial" here means primarily "linear"), namely non-linear, and only emerged in the second half of the twentieth century. It is considered the greatest achievement of classical physics in that century, and the associated mathematics is applied and developed in various other fields, such as in and with quantum dynamics.

There are always surprises on any sea surface, sometimes even surprising order effects such as standing or at least partially standing wave ("clapotis") when waves run against a beach, or even cross-shaped patterns when two wave systems from different directions overlap ("cross sea"). Figure 3.20 shows a clear crossing of two wave systems overlapping at the tip of an island.

The waves on a lake or the sea are a physical lesson for the world. They show the naked eye the cooperation of law and chance. In general, however, not much more happens in and on the world than the transition of kinetic energy into heat of the water. Occasionally there is a ship or boat accident and rarely, in the case of a major tsunami, a great catastrophe that robs us of our faith in the peacefulness of the sea and the most beautiful beaches.

Fig. 3.20 Cross lake. Two swell systems cross each other. Phare des Baleines, Ile de Ré. (Image: Wikimedia Commons, Michel Griffon)

In every moment, on every square metre or square centimetre of water surface, information is created and passes away. We will look at this process in more detail and more calmly below. As an introduction, let's take a closer look at the information that occurs.

3.2.3 What Information Is around us? Coincidence or Plan?

"Smooth figures are rare in the wild, but they are important in the ivory tower and in the factory."
Benoît Mandelbrot, French mathematician, 1924–2010.

The Subtle Difference
The ivory tower stands for classical science, the factory for what is industrially made by humans. We have already discussed this message of Benoît Mandelbrot in another form; now we examine the objects in wilderness and civilization a little deeper and see, on the one hand, objects whose blueprint is known and those whose blueprint is unknown or which we do not understand. The difference is fundamental and applies to the distinction in numbers, in programs, and correspondingly in the world at large.

With numbers we see in this sense on the one hand numbers with a (known) construction principle, for example the number 2/3, and on the other hand random numbers, where all digits are thrown: 5 6 1 3 2 as an example. Although the fraction 2/3 in the decimal system is an infinite sequence of "6", the calculation rule is available and simple. As an arithmetic program, it would be a few lines of code, and everything is fixed. It's different with random numbers or with numbers whose construction law you don't know or don't recognize: you have to communicate and explicitly store every single decimal digit.

The same applies to programs. If a running program is a black box, one must run it an unlimited number of times to grasp the program as a whole. If one knows the inner structure and the sequence of commands, one has a completely different, superior position. The length of the associated program is a measure of the complexity of the task the program solves. This definition once again highlights a paradoxical property of the Mandelbrot set:

Two simple equations, that is, a very short program, produce the most complicated object in mathematics!

The Mandelbrot set is at the same time one of the simplest and the most complex objects of mathematics, depending on the level of information, whether we know the equations or only the individual points and the graphs.

Information in the Physical World
In the physical domains, e.g. the lake and its waves or the sky with the clouds, chance dominates the physical construction laws that determine the type of waves or the type of

Fig. 3.21 An artificial "nature scene" with moonlight, clouds and rain. The physical model of the atmosphere used is NUMA, developed at the US Naval School. The visualization is done with the commercial software Maya®. (Image: Andreas Müller, from http://anmr.de/cloudwithmaya/. With kind permission)

clouds. But it is possible to create the external pseudofractal appearance of waves or clouds (or mountains) by synthetic chance and with software. The result is simulated images or models of the physical world, which themselves contain frozen randomness and thus appear realistic. Figure 3.21 shows an artificial "realistic", almost romantic scene, produced with today's technology. Although it is only a matter of outward appearance, the representation of clouds is a difficult task involving a great deal of physics: the shape of the cloud is pseudo-fractal, the light from the cloud is scattered several times until it reaches the viewer, and it is sunlight and light from the background of the sky. Whether real or virtual, the information of the cloud (and correspondingly of a water surface) has a physical core plus a lot of coincidence, generated from nature or in the computer.

Information in the Animate World

The biological domains of our environment, such as the trees and grasses, have another crucial source of data: by definition, they have a blueprint separate from physics, the genome. According to Wikipedia (pulled July 2020):

> **"The genome is the totality of the material carriers of the heritable information of a cell or a virus particle. In an abstract sense, it is also understood to be the totality of the heritable information (genes) of an individual."**

It is the rather stably stored software of life, which—together with the chemistry of proteins and amino acids as the processor—forms the basic framework for the structure of an individual.

Blueprints that are passed on to the next generation are proprietary knowledge and lessons learned from many previous lives (and deaths) and coincidences.

They are much more than the spontaneous formation of structures in the world of physics.

The diagram in Fig. 3.22 shows the sizes of the genomes of different classes of living organisms in the genetic unit of base pairs, i.e., pairs of amino acids that lie opposite each other in the helix of the double strand. Measured in normal information units, a base pair corresponds to just 2 bits; such dibits were also used in the past for the "modems" of communication via telephone lines.

First of all, we find a paradox, the so-called C-value paradox: the genome sizes do not correlate directly with the degree of complexity of the organisms. In animals, the amphibians have the largest genomes, while the largest genomes at all are found in plants,

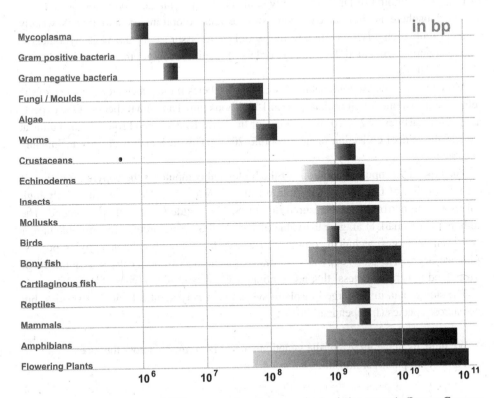

Fig. 3.22 The genome sizes of different organisms in base pairs (variation ranges). (Image: Genome sizes, Wikimedia Commons, Abizar at English Wikipedia)

and there in the Liliaceae or Liliales. As a measure of complexity, only the smallest program that executes the function sought can serve (Chaitin 2007).

The amount of information in the human genome is about 3.2 billion base pairs, or with 4 bases per one byte (i.e. one "octet" of bits) about 800 Mbytes per human and per cell. That's not a lot of information, and not a lot of required mass: the human blueprint is incredibly densely packed! The mass of the 3.2 billion base pairs is only about 3 picograms, or for normal cells doubled about 6 picograms. The number of cells with DNA nuclei in the body is highly speculative in the literature. The number of cells in the body is estimated to be 30 trillion (Greshko 2016). If we assume the order of magnitude of 10 trillion of these for cells with DNA, this would already mean the considerable mass of about 60 grams of DNA per human being, still without the DNA content of the 40 trillion bacteria living in us.

That's a surprisingly macroscopic amount! Even more astonishing is the amount of DNA in the biosphere. From the standpoint of information, the biosphere as a whole is a vast library of software blocks encoded in matter, in DNA. The authors Landenmark et al. (2015) estimate the amount of information in the world's biosphere to be 5×10^{31} million base pairs, or, in common units of measure in IT, about 10^{22} GigaBytes. This makes DNA almost a mass product: the bound mass is 50 billion tons, the volume about 30 million m^3 of DNA· The amount of DNA is roughly equivalent to the annual production of PET!

The biosphere is thus an enormous storage and computing system that is weakly coupled with each other—certainly not as tightly as assumed in the Gaia hypothesis.[3] The total number of stored data is many orders of magnitude larger than that of the largest technical data storage systems.

More tangible are the total data volumes of all species if one calculates the sums of the reference data alone for all probably 10 to 15 million animal and plant species. One expects about 20 PetaBytes or 2×10^7 GigaBytes for the Earth BioGenome Project, which aims at exactly this global collection (sangerinstitute 2018). This task size is already technically feasible today.

The typical "computer operation" of the biological computer is the copying function of the information of DNA, the transcription from DNA to RNA. Here, too, incredibly high values for the performance, measured in NOPS, Nucleotide Operations per Second. The authors Landenmark et al. give the value of 10^{39} NOPS, many orders of magnitude higher than even the largest technical computers. Even with a relatively reliable copying process with one error per 100,000 copying operations, this means astronomically large numbers of errors (and thus randomness) that are continuously introduced—a basis of evolution.

The stored data in the large biosphere system is largely identical within a species, but also across species (Independent, 2018):

- We humans, have 99.9% of the genetic material in common across the races,
- To the chimpanzees, it is still 98.5%.

---

[3] The Gaia hypothesis links the Earth's surface and the biosphere together into one organism.

- Cats still 90%,
- Mice 85%, domestic cattle 80%, fruit flies 61% and bananas 60%.

Genes and their environment are a multiple playing field of chance; superficially we see

- The very big, long-term game, evolution, that led to species,
- the random play within the species, the microevolution
- and current coincidences, in software language "in real time".
  medically termed "de novo" mutations,
- The mixture due to the random choice of partners.

We dedicate a separate chapter to the great game of chance "evolution with microevolution". Here for our analysis of chance we find that the blueprints of organisms have stored the coincidences of millions of years with a huge possibility of variation.

For real-time randomness, an example is the mutations that an individual accumulates and transmits, occasionally leading to, for example, trisomy, Down syndrome, or even schizophrenia.

In addition, there is the normal mixing of genes by chance during sexual reproduction. The human genome has at least ten million possibilities of variation in genetics:

"Two randomly selected, not closely related people differ in millions of base pairs as a result, estimated to be about four million base pairs each from a randomly selected other person."
From Wikipedia, "Genetic variation," pulled July 2020.

With these variations, many of our characteristics become random: body size, skin color, eye color, our susceptibility to various physical diseases, and some psychological traits.

The genetic differences become "eye-catching" in the truest sense of the word in eye colour and generally in the colour structure of the human iris, the aperture of the eye (Fig. 3.23). Several genes are involved in colour formation. If you look a little more closely at the eye, you will notice details in the color pattern of the iris. These patterns form (on a genetic basis) in the first few months of life and then remain virtually unchanged for the rest of life. They are typical of the individual and are not identical even in identical twins. This is the basis for the use of iris recognition as a simple and precise method of identifying people, biometrics.

In general, the biometric methods used are examples of "marked randomness in the individual" or of great disorder in the order of the living human being. Such random effects are, for example, the lines on fingers and on the hand, the shape of the ears, the fine structure of the voice and finally the DNA or the "genetic fingerprint". For practical application as access control or in criminology, the criterion must be as unambiguous as possible in the whole of humanity, be consistent for an individual throughout life, be present in everyone if possible, and in addition be easy to measure.

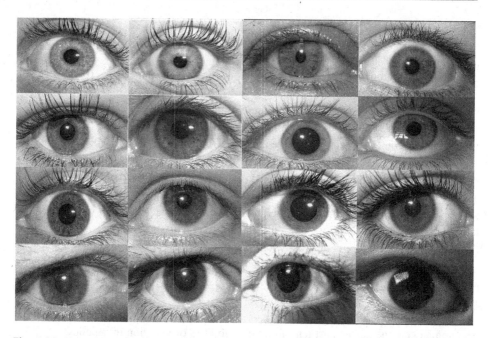

Fig. 3.23 Eye and genetic chance: the variation of eye color. (Image: Color gradient eye colors, Wikimedia Commons, LeuschteLampe)

A fundamental mechanism for the generation of typical macroscopic random patterns in nature was found and mathematically described by the brilliant computer scientist Alan Turing in 1952. It was Turing's third great original idea after the Turing machine (the computer model of theoretical computer science) and the Turing test (as a pioneering contribution to artificial intelligence). Several decades were to pass before the idea of this "Turing mechanism" was recognized.

In the chemical interpretation of the mechanism, two substances that are in themselves homogeneously distributed act together (or better against each other) and produce stripes, spots, spirals or hexagonal structures. From the small random (sub-) microscopic fluctuations, typical macroscopic random patterns are thus created, which often constitute for us the beauty of a plant or an animal. According to Popper, we call this "directed randomness". Figure 3.24 shows the skin of the gold-ringed puffer fish as an example of such a Turing pattern, but of course zebra and cheetah fur are also wonderful examples from biology.

A physical example are the ripples in the sand of the dunes (Fig. 3.25), which can also be seen as Turing patterns. The wind builds up ridges of sand. In this process, the ridge acts as a sand depression and reduces the flow of sand in the air behind it; this results in an approximately equal distance to the next sand wave in the dune. For comparison, the conditions for water waves—although non-linear—are clearer: there it is a matter of the dynamic, quite random distribution of the various forms of energy involved, such as surface tension, gravitational and kinetic energy among themselves.

Fig. 3.24 Turing pattern on the
skin of the goldring puffer fish.
(Image: Giant Puffer Fish Skin
(detail), Wikimedia Commons,
Chiswick Chapter)

Fig. 3.25 Turing pattern in the sand of a dune in Erg Chebbi in Marollo. (Image: Morocco Africa
Flickr Rosino (detail), Wikimedia Commons, Rosino)

We see: Under the smooth lawful structures, for example of eye, skin or fur, is
coincidence, very much coincidence, measurable in bytes as information. Despite the
basis of the plans, there is chance everywhere: in the meadow with the blades of grass
and flowers, in the forest with trees and leaves. The closer you look; the more chance you
will discover. The blueprint for the large oak tree comprises about 185 megabytes, but the

information needed to capture the large, fully grown tree, branch by branch, leaf by leaf, even in the details that are still visible to the naked eye, is orders of magnitude larger.

Most of the coincidence seems harmless, but overall we do lose security with it. And even just the color of skin or hair can change a life:

> **"The blonde jokes .... should never be taken seriously. Anyone who thinks hair color (or skin color) or any other outward characteristic is an indication of a certain character or of special intelligence is usually lacking in that respect himself."**
> **Foreword to "Blonde Jokes" in Programmwechsel.de**

3.2.4 Summary of the Chapter

All around us is chance. We do not want to overlook it or average over it, but to appreciate it, although ignoring or almost ignoring it has been so successful in science and meant a credo for classical art, which until the nineteenth century loved above all the smooth, perfect. The first mathematician to study disordered, curly nature with chance was Benoît Mandelbrot. His mathematical method for this is the geometric theory of fractals. By this theory, the curled curves of nature acquire an effective dimension greater than one, and the folded and rough surfaces acquire a dimension greater than two.

For Mandelbrot, the mechanism by which nature grows is self-similarity, the repetition of a construction in a smaller (or larger) size. In mathematics, of course, this goes to infinity, in nature only over two or three orders of magnitude, then the physical boundary conditions with which nature solves a task change, such as "providing as much surface area as possible with as little mass as possible". Such "fractalesques" can be found everywhere, whether in cauliflowers or snow crystals.

If we look around us in everyday life, we discover coincidence everywhere. If they are industrially produced smooth shapes, we have to take the microscope or go down to nanotechnology.

A particularly beautiful example of chance and regularity are waves on a lake or the ocean. The spectrum ranges from light chance and lovely waves to erratic monster waves with deadly chance. When the wind fails, the waves sink and their energy and information disappears in the thermal tremors of the water molecules.

> **Waves can be used to understand how randomness works in nature: It is the interplay between different physical forces that build up or dampen randomness.**

In the world of the living there is something fundamentally new, namely blueprints. The blueprints themselves are "frozen chance". Chance plays with the blueprints, it changes them physically slowly or moderately fast (evolution and microevolution) or it mixes them anew, for example in sexual reproduction. In any case and everywhere we have a lot of chance in the appearance, whether in humans or in a tree. Beneath this is massively hidden information in individual blueprints, and beneath (or above) all this are the natural laws of physics as a foundation. It is the intention of the book to show that chance is built into the foundation.

Most of the time, little chance in the world is just a kind of passive noise. But isn't it disturbing, after all, that chance is at work almost everywhere in the world? It is almost a hidden pantheism!

References

Braden, Gregg. 2009. *Fractal time: The secret of 2012 and a new world age.* Carlsbad: Hay House.

Brookman, W.R. 2015. *Orange proverbs and purple parables.* Eugene: Wood & Stock.

Chaitin, Gregory. 2007. *Algorithmic information theory – Some recollections.* arxiv.org/pdf/math/0701164.pdf. Zugegriffen im Juni 2020.

Greshko, Michael. 2016. How many cells are in the human body – And how many microbes? Nationalgeographic.com/news/2016/01/160111. Zugegriffen im Juni 2020.

Hehl, Walter. 2018. *Galileo Galilei kontrovers.* Heidelberg: Springer.

Landenmark, Hanna, et al. 2015. An estimate of the total DNA in the biosphere. *PLoS Biol* 13 (6): e1002168. https://doi.org/10.1371/journal.pbio.1002168.

Sturm, Hermann. 2003. *Alltag und Kult: Gottfried Semper, Richard Wagner, Friedrich Theodor Vischer.* Basel: Birkhäuser.

Turing, Alan. 1952. The chemical basis of morphogenesis. *Philosophical Transactions of the Royal Society* 237: 37–72.

Understanding Zufall (Coincidences) in the World

"The eternal mystery of the world is its comprehensibility. This fact is a miracle." (Albert Einstein, German-Swiss physicist, 1936).

We begin the attempt to understand chance in the world with the beginning of the cosmos, the Big Bang, from a quantum fluctuation. This begins an incomprehensible sequence of coincidences or necessities that lead to our existence, to the stardust from which we are made, and to our habitable earth. Here we see the story of our earth and solar system so fitting for us, it is tailor-made. It is the emotionally so seductive, almost meaningless anthropic principle.

Einstein refers to the status of physics and, in particular, the progress in modeling the cosmos on the basis of "his" theory of general relativity. In the decade before this quote, the basic construction of the universe became visible—the distance of the Andromeda nebula M31 was determined, still too small but still far outside the Milky Way. The redshift of galaxies had been known since 1912, and in 1927 the Belgian priest Georges Lemaître had already deduced from it the expansion of the universe. His work in French had appeared in a very special journal, the Annals of the Scientific Society of Brussels, and remained almost unknown, although Einstein promoted Lemaître and his thoughts after initial hesitation.

In 1929, the American observational astronomer Edwin Hubble published the relationship that is now officially called the Hubble-Lemaître law: The farther away a galaxy, the greater the redshift. It was first thought to be a common Doppler shift due to the removal of the objects, but today we know

- space itself expands between galaxies, and
- there was a beginning to the universe.

Lemaître is considered the discoverer of the "Big Bang", the Big Bang, which he called the "primordial atom", the uratom. The word "Big Bang" had been invented by British

© Springer Fachmedien Wiesbaden GmbH, part of Springer Nature 2021
W. Hehl, *Chance in Physics, Computer Science and Philosophy*, Die blaue Stunde der Informatik, https://doi.org/10.1007/978-3-658-35112-0_4

astronomer Fred Hoyle as recently as 1949—to mock the idea. Hoyle had thus created one of the most popular scientific terms.

The Big Bang had also triggered religious thoughts about the classical creation by a God, but this is inappropriate: Either God is a principle, in which case it is just an additional word for a gap in knowledge that explains nothing, or God has human features, in which case the image is totally inappropriate for the act of creating the cosmos with matter and energy, space and time. It is not about the creation of the world in time and space, but about *the* creation of time and space.

What Einstein begins to understand, but does not accept, are the beginnings of quantum theory. Einstein and the physicists of the first half of the nineteenth century, however, with relativity and quantum theory, only have the understanding of the physical side of the world, but there is more. The whole information processing side is missing. It is only from about 1950 (a few years before Einstein's death) that it becomes apparent that the "mental" side of the world has something to do with emerging information processing. It is a matter of understanding in principle the questions of Emil du Bois: How does one feel? How does one think? It is great and could not have been foreseen in 1872 as in 1936 that we would be able to answer these questions. The world model should therefore include not only physics, but also the biological and intellectual side of the world and, of course, chance. And chance begins with the creation of the universe.

4.1   Big Bang and the First and Second Coincidence

"We people of the earth exist because the potential for our existence was laid in the Big Bang, 13.7 billion years ago, when the universe exploded into being."
Robert Brown, Australian author and politician, born 1944.

4.1.1   The Emergence of the World from a Vacuum

The vacuum is not a nothing or the void, but a

"boiling, bubbling brew of virtual particles that appear and disappear, but so quickly that we can't even see them directly."
Lawrence M. Krauss, American physicist, born 1954.

The vacuum of the universe breaks open—by chance, and there has been the big bang. But this is "postmodern physics" in our parlance: Physics with many alternatives and theories awaiting confirmation by experiment and observation. What is concretely present is

- the measured expansion of space according to Hubble-Lemaître,
- the cosmic background radiation, discovered by Arno Penzias and Robert Wilson in 1964 with an experimental antenna at Bell Labs.

In addition, there is an immediate problem with the formation: For reasons of symmetry, matter and antimatter should be created in exactly the same mass—but we live in a universe of normal matter; every lump of antimatter in our world immediately radiates to a maximum of possible energy. It is only now, in 2020, that experimental evidence is emerging that certain particles, the easy-to-create and elusive neutrinos and their opposites, the antineutrinos, do not behave fully symmetrically.

Here are some recent headlines (April 2020):

- *"The Yin-Yang of the Big Bang" (World Science Festival).*
- *"Why the Big Bang Did More Than Nothing" (NY Times).*
- *"Neutrinos explain why we don't live in a world made of antimatter" (New Scientist).*

This almost random-looking difference could be the key to the whole existence!

The cosmic background radiation in the microwave range was immediately identified as the expected aftermath of the Big Bang. The radiation is, in the jargon, still the "smoke of the fired gun" and stems from the time when the universe had cooled down just enough to form neutral atoms—300,000 years had passed since the Big Bang. The cooling by a factor of 1000 produced the background radiation, corresponding to quite exactly 2.725 Kelvin residual temperature, which is the same in all directions with only minor spatial residual random fluctuations.

Thus the universe arose from a coincidence (or two coincidences, if one considers the matter-antimatter problem to have been solved by chance): From the Big Bang with a random preponderance of matter over antimatter in a quantum fluctuation (Fig. 4.1), when even space and time were a kind of particle or "bubble". The great variation of fluctuations of the beginning is followed by a curious period of mixing and a curious, extremely short

Fig. 4.1 The hypothetical chaos of quantum fluctuations at the Big Bang. Artist's illustration of the "quantum foam". Image: NASA/CXC/M. Weiss

, and extremely large expansion, inflation, which pulls the originally neighboring points far apart and produces the homogeneous radiation pattern of today. After a few hundred million years, the remaining random density fluctuations intensify into relatively denser gas clouds, from which the first stars form after about 200 to 400 million years, large, young stars. The supernovae of these stars then give rise to the stardust of which we happen to be made:

> "The nitrogen in our DNA, the calcium in our teeth, the iron in our blood, the carbon in our apple pies - we are made from the insides of collapsing stars. We are made of star stuff."
> Carl Sagan, American astronomer and author, 1934-1996.

Of course, the carbon in our protein is actually made of stardust too, not just apple pie.

4.1.2 Anthropic Principle and Goldilock Puzzle, Necessity or Agglomeration of Coincidences?

> "The Goldilocks enigma is the idea that everything in the universe just fits for life, like the oatmeal in the story."
> Paul Davies, British physicist and author, born 1946.

Origin is the English children's story from the nineteenth century about the girl Goldilocks. In the usual version, the girl Goldilocks has invaded the cottage of the three absent bears and eats of their porridge, sits on their chairs and lies down in their beds. But there are three porridge bowls, three chairs and three beds—she tries them all and only the last one ever fits her. In this sense we live on a Goldilocks planet where everything is (or was) optimal.

The universe, from the beginning to the present, is also a work of chance or coincidence. But it all fits together beautifully: Our sun and its mass and chemical composition, allowing stable nuclear fusion over billions of years, and the composition of the earth and our bodies, allowing carbon-based life to emerge, with green chlorophyll to match the light of the sun with magnesium and red hemoglobin for oxygen with iron.

The more genteel, philosophical version of the Goldilocks is the Anthropic Principle, formulated 50 years ago, but known as a basic idea for longer than the sense that the universe must be as it is in order to produce man (Wallace 1904). The term "anthropic" (from the Greek *anthrōpikos*—'of or for man') shows the first-person nature of us humans: Everything is made so that we exist. This statement cannot be hard refuted according to today's knowledge—the contradiction and the thesis that there are an infinite number of worlds cost Giordano Bruno his life in 1600! The sentence *Everything is made so that we exist* can be interpreted in different ways, "neutral" or "egoistic-human":

- Neutral or "weak": It's just the way it is. Period.
- Human-causal or "strong": "Someone" has just made this so for us!

This emotional and non-scientific discussion about the "weak" or "strong" anthropic principle can be particularly well conducted by quoting the so witty German physicist Georg Christoph Lichtenberg (1742–1799):

> **"He [the philosopher] wondered that the cats had two holes cut in their fur just at the place where they had eyes."**

Here, however, we are observers from the outside (though not "creators"):

We can only "weakly" accept the seeing cat eyes as fact, or go deeper and "strongly" understand that they arose in evolution together—eye, eyelids and fur. Proof that the statement is not trivial would be the observation of bugs: For example, kittens that still have fur over the eye after birth. There is even the "third eyelid" in cats, the nictitating membrane, which can really protrude in some diseases. In principle, even a sighted cat could notice that some cats have a disorder.

With the universe we do not stand by, have no comparison, and the existence of many universes is only a postmodern hypothesis. Rationally, there is the weak proposition alone, but there are strong feelings (on this, see Hehl 2019): the list of anthropic coincidences, the effects that make our very existence possible, is long, several dozen phenomena long. It starts with the fine-tuning of natural constants, the subtleties of the Big Bang, then the evolution of the Sun, the emergence of the Earth, and the Earth's position in the solar system's "habitable zone" (Fig. 4.2).

Probably even the unusually large earth moon is a stabilizing element for our development. It really looks emotionally as if everything is artificially created, "designed" for us. But perhaps several dozen independent coincidences are not necessary at all: Much or

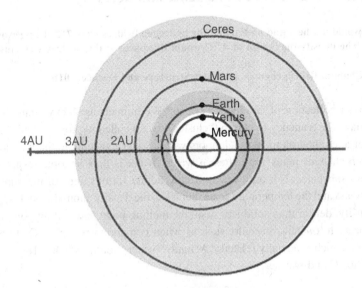

Fig. 4.2 The habitable zone in our solar system. Image: Wikimedia Commons, EvenGreenerFish

even everything is connected. Evolution could not start earlier than 4 billion years ago (the Earth was formed about 4.5 billion years ago), evolution took the minimum 3.5 billion evolutionary time, shorter was not possible, and carbon-based life must be. There is no other chemical basis. For example, the related element silicon does not work as a basis for life. Only carbon chemistry is flexible enough for life. This is not "carbon chauvinism", but chemistry under the fairly safe assumption that the periodic system applies throughout the universe and there is nothing better, and nothing else.

The situation is different for energy generation and storage. Here, a wide variety of process variants are conceivable, e.g. on the basis of sulphur or hydrogen sulphide (see below).

The aforementioned British physicist Paul Davies (b. 1946) writes:

"There is something going on behind it all . . . the answer is to be found within nature, not beyond it. The universe might indeed be a fix, but if so, it has fixed itself."

Then there would be much less such big coincidences than intuitively assumed. But for that we probably need to understand more of the basics.

It goes right on with tinkering and coincidences: The water is also a great physicochemical tinkering in our favour.

4.2   Water and Chance

4.2.1   Water Properties as Coincidence and Necessity

"Water would not be liquid on Earth if the hydrogen bridges were 7% stronger or 29% weaker. The density maximum at 4 °C would disappear if the bridges were just 2 % weaker".
Martin Chaplin, British chemist, in Water Structure and Science, 2018.

This is another collection of Goldilocks effects! We humans are lucky again. The water molecules have the tendency to clump ("cluster") by so-called hydrogen bonds in the four directions of a tetrahedron to four neighbors with a water molecule in the middle.

This lump changes many properties of water. Water is thus in some respects a quite extraordinary substance. Not only was it used to define various units of measure, such as the unit for mass and the temperature scale, but it has the density anomaly: At 4 °C it has the greatest density, denser than solid ice. Also, the melting point and boiling point are much too high for such a small molecule, such as when comparing water (H_2O) to hydrogen sulfide (H_2S), which is actually related. "Actually" water should melt at -100 °C and be a gas from -80 °C and warmer.

These hydrogen bonds also prevent water molecules from displacing each other too easily. This resistance is called viscosity; viscosity is of fundamental importance for biology and physics. In biology, viscosity is crucial for the passage of membranes and for cell division; in physics, it is crucial for the flow through tubes and the generation of waves.

For us, viscosity is a measure of the emergence of randomness: the higher the viscosity, the less randomness. With the phenomenon of turbulence in a fluid, the randomness and its emergence becomes directly visible.

4.2.2 Water and Vortices

Do water vortices always turn in the same direction when flowing down the bathtub in the northern hemisphere? The answer is: No!

Before we look at our domestic vortex, a note in resonance with our cosmic thoughts in the previous chapter: The phenomenon of the vortex exists on a wide variety of scales in nature, from microscopic vortices to vortices of tornadoes to the vortices of galaxies.

The universality of the concept of "vortex" is demonstrated by the two examples in Fig. 4.3: The familiar bathtub drainage vortex on the one hand, and the spiral galaxy M51 or NGC 5194 or Whirlpool Galaxy on the other. Commenting on the Hubble Space Telescope image of the galaxy M51, the usually sober NASA says: "The graceful winding arms of the majestic spiral galaxy M51 appear to be a great spiral staircase sweeping through space." The diameter of this Milky Way is 80,000 light years, so the difference in scale between the images is about $1: 10^{20}$! But the basic physical principle and the

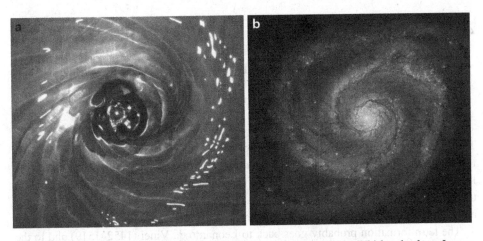

Fig. 4.3 (a) Bathtub vortex. Image: Flickr/Pete Keogh. (b) Messier51, the Whirlpool galaxy. Image: Wikimedia Commons, **ESA/Hubble S. Beckwith**

beginning are the same: The basic principle of vortices is the conservation of angular momentum, the momentum in the initial matter.

The question at the beginning of the section is also often asked in the form:

Does the bathtub vortex spin differently in the southern hemisphere than in the northern hemisphere of the earth?

The occasion for the question is due to quite a few (incorrect) popularizations of the Coriolis force, the apparent force effect on moving bodies on the rotating globe. In a 1995 episode of The Simpsons, the effect plays a major role, according to the Wikipedia article 'Bart vs. Australia', pulled July 2020:

"Bart notices that the water in the bathroom always drains counterclockwise. Lisa explains to him that this is only the case in the southern hemisphere, because of the Coriolis effect."

This is physically correct in principle, but in practice it is wrong. At rotating earth-sphere at northern hemisphere a movement towards a depression all times is deflected to right side and thus results counter clock-wise rotation, south of equator however to left side and results clock-wise rotation. This also applies to low-pressure areas. At equator there is no such deflection at all. The problem is the smallness of the Coriolis force (or the smallness of the bathtub): The random vortices within the water and the shape of the tub are dominant— one must have a larger circular basin with a central drain, wait 24 hours without any air movement until the internal vortices have subsided, then one has a chance to observe the "real" vortex. Usually the direction of rotation of the outflow vortex at home is a nice example of "random" chance, in the north as in the south: the time of onset and the direction are random.

4.2.3 Turbulence

"When I meet God, I will ask him two questions: Why relativity? And why turbulence? I already believe he has an answer to the first question."
John von Neumann or Werner Heisenberg attributed, probably by the British physicist Horace Lamb, 1849–1934.

Turbulence is one of the most difficult areas of physics including the associated mathematics. Blame is put on chance, which breaks in through turbulence, its outbreak and its effects in turbulent flows. On the other hand, next to turbulence, in the same water, there is order in the form of calm flow; the transition between turbulent and calm is the central problem. For an analysis of chance, turbulence, eddy or not, is a visual lesson.

The term formation probably goes back to Leonardo da Vinci (1452–1519) and to the Latin *turbare*—'to twist, to tangle'. Leonardo da Vinci was fascinated by water and painted a series of water pictures with swirls like locks of hair (Fig. 4.4) and wrote:

Fig. 4.4 Water studies by Leonardo da Vinci. Image: Old Man with Water Studies (detail), Wikimedia Commons, drawingsofleonardo.org

"The rippling motion of the water surface is similar to the behavior of hair, which has two motions, one of which depends on the weight of the curl and the other on the direction in which they turn – that's how water forms eddies."

Leonardo da Vinci takes his paintings to the limits of signability, with his delicate curls over curls in water or the faint outlines of streams of water (and of explosions).

In the pictures (Figs. 4.4 and 4.5), vortex streets develop behind obstacles in an otherwise smooth, "laminar" flow. Turbulence also arises from tiny beginnings, grows or

Fig. 4.5 Water flow around obstacle. Image: Turbulence, Wikimedia Commons, aarchiba at English Wikipedia

disappears again. In turbulence, kinetic energy is constantly required to maintain the motion.

The already mentioned pioneer of meteorology Lewis Fry Richardson found the vortices so important that he wrote a famous "physical" verse about them:

"Big whirls have little whirls, that feed on their velocity, and little whirls have lesser whirls and so on to viscosity."

Roughly translated:

"Big vortices have little vortices feeding them with their velocity, and little vortices have smaller vortices, and on and on until viscosity."

Richardson invented the verse after a poem by the British mathematician Augustus De Morgan (1806–1871), who in his verse *Siphonapter a* (flies) linked "large flies" with small and smaller flies, and upwards even larger flies *ad infinitum*. Both "poems" anticipate the idea of fractals, fractals for vortices on vortices, flies on flies, downwards but also upwards to the galaxies.

How vortices are created—and thus randomness and chaos from order—was investigated by another British physicist, Osborne Reynolds (1842–1912), with experiments using coloured water. The basic principle of emergence is already shown by a flag in the wind (Fig. 4.6). The plume is an unstable boundary surface that bisects the flowing air.

Whirls are created, alternating left and right, which run irregularly through the flag and bend the fabric surface in the process. At the end of the flag, the vortices separate as a slightly chaotic "vortex street": A vortex on the left, a vortex on the right, and so on. In strong winds, the fabric of the flag even rattles. A flag without air and in vacuum—famous is the flag on the moon—moves totally different: The flag swings only after touching, for example after standing up, and then slower, more powerless and much longer than on earth. But there is still no reason to doubt the moon landings.

However, such neighbouring layers with some other properties, e.g. slightly different speeds, practically exist all times, even without separating substance, and small vortices come up, which can become larger. So waves come up on water and vortices come up in air.

The decisive factor for the formation of vortices is the ratio of the velocity energy to the viscous damping, the motion to the viscosity of the medium.

An excess of kinetic energy favors the formation, the toughness slows it down. In addition, the size of the object. The longer the typical path for the possible build-up of vortices, the greater its probability. The ratio is called the Reynolds number according to the physicist mentioned above:

Above a certain Reynolds number, measured or calculated at a point in the flow, the previously smooth filaments of the flow become tangled and vortices can form and grow, and the calm laminar flow pattern ("orderly") changes to turbulent ("random" or "chaotic", Fig. 4.7). The resistance of the flow to the movement increases abruptly.

An example are tubular pipes: With small pipe cross-sections, the soothing viscosity predominates, with larger ones the hectic kinetic energy—the transition point depends (calculably!) on the velocity and on the type of liquid or gas.

This raises the fundamental questions:

Fig. 4.6 A flag in the wind as a device for creating vortices. Image: Swiss Flag in the grounds of Auberge du Lion d'Or. Wikimedia Commons, smuconlaw

- Where does the very first push and start come from?
- How much order is left in the disorder? What is still valid?

But also as a third question:

- How does a disturbance or disorder fade back into order?

4.2.4   Between Chances and Order

"πάντα ροιζει panta rhoizei—everything rushes".

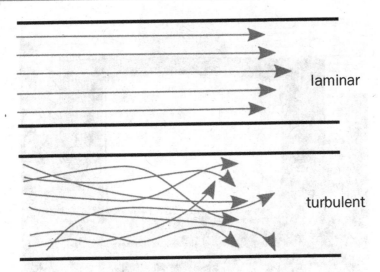

Fig. 4.7 The two flow types laminar (straight flow lines) and turbulent (swirled lines). Image: Laminar and turbulent flows, Wikimedia Commons turbulent Dubaj/Guillaume Paumier

Saying modelled on the formula "everything flows" by the Greek philosopher Heraclitus.

The First Kick-Off or: Everything Rushes

There is no stillness or silence in the world, at least not at room temperature and not in matter. Our hearing would only have to be a little more sensitive and we would feel the pattering of the air molecules on our eardrums. But that would not have been useful information for life and evolution! We have already described above how it became clear in the nineteenth century that heat means mechanical random motion. The movement or trembling or oscillation takes place "deep down there", at the molecular level, but occasionally it is still indirectly visible, e.g. in a light microscope as Brownian motion.

With the advent of electronic amplifiers at the beginning of the twentieth century, electronic noise became measurable, audible and visible. In 1918, the German physicist and electrical engineer Walter Schottky described the first noise as measurable irregular current fluctuations. It is the statistical fluctuations of electrons when leaving a glowing cathode, when crossing a barrier or simply the thermal movements of the electrons. In the loudspeaker, this results in the typical noise that gave the phenomenon its name: noise. Noise is one big random process: Called "white" when the subordinate coincidences have similar energy even at higher frequency, "pink" when higher energies are less represented. The names are taken from the colours of light: "White" light contains all colours in a certain sense, "pink" prefers long wavelengths with less energy. White noise is "the most random thing there is". If you encode it in numbers, there is by definition no way to predict the next number! You have to wait until you get the new information. It's a stream of maximally disordered technical information, and thus a stream of maximal entropy—but completely

Fig. 4.8 Visible coincidence: An analog TV screen without a program signal and therefore with noise. Image: TV_Noise, Wikimedia Commons, Mysid

worthless for interpretation. Unless, of course, one needs just that! A hardly noticed pretty paradox: Maximum entropy and information with maximum meaninglessness.

The difference in "listening to chance" is striking: White noise is unpleasant and piercing, pink noise is rather friendly.

Coincidence can also be visible; especially beautiful (and audible) in old, analogue television sets that are switched on without receiving a station (Fig. 4.8). Then one immediately sees the dynamics of chance that surrounds us everywhere. There is noise everywhere and thus chance in miniature defined as:

▶ **Definition Noise is an irregularly fluctuating quantity with a broad frequency spectrum.**

The list of noise phenomena is long: The radio background of the cosmos, the discharges in the atmosphere, the seismic vibrations in the earth's subsurface, Brownian motion, thermal noise, noise in ferromagnets, and so on. In a broader sense, the variety of shapes of leaves on a tree or the roughness of a wall is also a kind of noise.

In modification of the saying of Heraclitus πάντα ρεῖ **panta rei—everything flows,** which has already been mentioned several times, one can say:

πάντα ροιζει panta rhoizei - everything rushes.

Thus it is clear how in a water pipe the turbulence takes the beginning: By an elevation in the wall, a density or a velocity fluctuation. Then with the turbulence the coincidence grows. There is randomness everywhere.

The Boundary Conditions of Chance

> "People do not understand how what is borne in different directions comes to be in agreement with itself."
> Heraclitus, Greek philosopher, 550 - 460 BC.

We have already seen (and constructed) situations in which arbitrarily large events have arisen from the microcoincidence of "butterfly wings flapping". But the common, everyday coincidence constantly flows around us as incessantly as the current of the mountain stream in Fig. 4.9.

A roughly constant amount of water per unit of time flows past the observer over a long period of time. Laminar currents and turbulences, even whirlpools alternate. Wood chips or paint additives brought into the water show the path of the water particles, as straight-line motion or straight threads in the laminar part, and then the chaotic swirling in areas of turbulence. However, as a whole, for nearly the whole amount of water is valid: The water reaches the valley. However, position of single water-particles within outlet of stream is undetermined and appears as a coincidence.

Fig. 4.9 A mountain stream with laminar and turbulent flow. Oil painting by Paul Weber (1823–1916). Image: Wikimedia Commons, basenge.com

This alternation of uniform progress and random whirl is a familiar model for the course of history with a general direction ("the progress") and phases of revolution with inherent randomness:

"What if we believe that there is an overwhelming amount of randomness and path dependence [with chance] in history?" (Stanford Encyclopedia of Philosophy, article "Philosophy of History")

In the German philosopher Immanuel Kant, in his customary codified form of expression, we find a similar thought on world history from the year 1784:

"... it [history] can discover a regular course of the same; and that in this way, what in **individual subjects is convoluted and irregular to the eye, can nevertheless be recognized in the whole species as a steadily proceeding, although slow development of the original predispositions of the same."**

Kant thus sees again and again turbulent phases, even destructive periods, but on the whole—in the favourable "moral" case—a positive overall development.

In a mountain stream, the laws of physics provide internal boundary conditions (in addition to the external boundary condition of the rocks of the streambed). The law of conservation of mass applies, except for splashes of water from the side, and even conservation of volume, since the water is almost incompressible. Plus the laws of conservation of momentum and angular momentum and the law of energy. The metaphors for the course of history show a difference from physics. There is, for instance in Kant, the idea that there is a guiding principle of history and that the end point (he says the "intention of nature") is somehow good. This is teleological thinking; physics (just like biological evolution) says nothing about going on. But physics keeps chance in check in these examples.

We mention two more classical philosophical concepts in the environment of chance, which can be clarified with the types of flows, laminar and turbulent: Determinism and Indeterminism. Classically, (causal) determinism is understood to mean that all processes are causally determined and can be traced back arbitrarily far. The processes in the determinate world are like smoothly flowing streams. In contrast, indeterminism is the doctrine that an event came into being without a cause—a thoughtless absurdity for us humans. It would all be turbulence. For a limited situation, such as the flow of the stream between two points A and B, one can clearly define:

- If there is a streamline between point A (top) and B (bottom), then the flow between A and B is deterministic.
- If no flow thread can be traced from A to B, the flow between A and B is indeterministic, i.e. cannot be determined.

However, neither of these classical terms is generally applicable in any meaningful way.

The Decay of Chance

"Il mare poco a poco si calma - very gradually the sea calms."
from "Idomeneo or the Calming of the Sea"
Opera by Wolfgang Amadeus Mozart 1781.

We create a model coincidence at the lake in the lake water: The author jumps into the lake (Fig. 4.10). From "the lake's point of view" it is a coincidence, because there is no connection between the lake and my brain. True, Richard Feynman, the famous physicist, says "water" is the worst possible example for beginners in physics (Feynman 1963), because there are different kinds of waves, capillary waves and gravitational waves. But for our philosophical purposes it is well suited and everybody knows the behavior of water. However, Feynman is right, especially for the first phase of the jump, when the jumper creates the so-called near field in the water. This process is complex.

During the jump (Fig. 4.10a) the body in its special shape pushes the water to the side and creates the hectic near field (Fig. 4.10b). Subsequently, the opened channel collapses with momentum, waves part again, rejoin. It is a "small bang", a small or better a very small bang. The inner details of this bustle are determined by the geometry of the body, the "source". But in the implosion, they are annihilated by the back and forth of energy in a confined space and the mixing vortices of water.

We will find the same basic idea of the annihilation of information by implosion in a completely different setting, in the bottleneck effect of evolution.

One or more circular waves emanate from the point of entry (Fig. 4.10c). Only very gross asymmetries of an incident body distort the circles, at least at the beginning, such as a long stick or a cuboid thrown lengthwise into the water.

The near field quickly disappears and superficial calm occurs at the point of entry. The rings of circular waves—the far field—spread out over the lake. It comes to the last phase: The wave amplitude becomes smaller and smaller, the energy of the jump spreads until it becomes unidentifiable in the more or less calm lake surface. In the limiting case of the calm surface, it is lost in the thermal motion of the molecules of the water surface. Figure 4.11 shows the lake surface with centimetre-high random waves sparkling in the sun, which disappear, move or are rebuilt after only one or two seconds. The lake is thus a macroscopic random machinery, about ten million times coarser than the foundation of thermal fluctuations.

The leap into the water is a trivial example, but well understood and philosophically hard to overestimate: The collapse of a coincidence, the creation and destruction of information, the annihilation of the individual past, the spread of the brief news of the event, and finally the sinking into the anonymous fluctuations of the underground.

Even the largest supercomputer is then no longer able to reconstruct a single sunken event like this jump. The causal chain, starting with the jump, has disappeared in the noise. Everything has been, in principle, computable and deterministic—but undetectable. In the

a) Immersion in the lake. Beginning of the disturbance **b)** Visible near field

c) Beginning far field.

Fig. 4.10 The phases of the emergence of waves from a point source. Emergence and decay of randomness using the example of a jump into the water. (**a**) Plunge into the lake. Onset of disturbance (**b**) Visible near field. (**c**) Onset far field. Pictures: Edith Geissmann

section on the philosophical significance of noise, we return to the example of the lake as a random machine.

4.2.5 Summary of the Chapter

We begin the attempt to understand chance in the world with the beginning of the cosmos, the Big Bang, from a quantum fluctuation. This begins an incomprehensible sequence of coincidences or necessities that lead to our existence, to the stardust from which we are made, and to our habitable earth. Here we see the story of our earth and solar system so fitting for us, it is tailor-made. It is the emotionally so seductive, almost meaningless

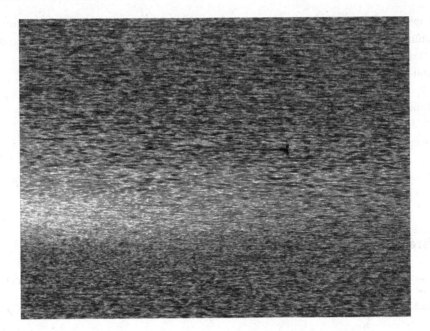

Fig. 4.11 A lake surface with myriads of random capillary gravity waves. Image: own

anthropic principle. It is especially true of the properties of water, e.g. that water expands when it freezes and is liquid at all at our usual temperature! A tiny change in physics, and everything would be different.

We take water and our experiences with water as a lesson in chance. Water as a fluid with low viscosity shows both, order and chaos or chance. In particular, we use the example of water to understand a universal dynamic random phenomenon: Vortices. It is similar from distant galaxies to bathtub outflow across 20 orders of magnitude. For this purpose we study (like Leonardo da Vinci) vortices and their origin from smallest beginnings and show, causes exist nearby everywhere, cause vortices are rushing nearby everywhere, lastly also all times "at bottom", at atoms, molecules, electrons or quanta. Often conditions are given, small coincidence becomes more and more coincidence—however not unlimited. Mostly physics limits, because the theorems of physics are valid e.g. for momentum and energy. In addition, we note that there are two philosophical possibilities in the world almost side by side: Ordered processes ("deterministic") and chaotic ("indeterministic").

We consider the throwing of a stone or the jump into the water to be particularly instructive. **For the lake, a coincidence (the Zufall) breaks in with it. The impact creates a chaos that mixes everything in a limited area and destroys a large part of the past.**

A long-distance wave spreads out, becomes increasingly fainter until the event sinks irretrievably into the visible or microscopic noise of the water surface.

Implosion (as after immersion) is identified as a general mechanism that annihilates structure and information. One speculation here is to view the primordial cell of the cosmic Big Bang as such an implosion cell that obliterated all that came before.

Thus we see the general random processes at the water waves:

- Building chance from tiny microscopic beginnings,
- Destruction of information by implosion,
- Decay of randomness and return to the microscopic (noise).

A lake with wind or the sea make it visible with their waves: Dynamic randomness is everywhere.

References

Feynman, Richard. 1963. *The Feynman lectures on physics*. Vol. I. Amsterdam: Addison-Wesley.
Hehl, Walter. 2019. *Gott kontrovers*. Zürich: Vdf.
Wallace, Alfred Russell. 1904. *Man's place in the universe*. New York: McLure, Phillips & Co.

Three Worlds in the World, with Coincidence 5

> "The universe is full of magical things waiting patiently for our wits to grow sharper."
> (Eden Phillpotts, English writer, 1862–1960).

The historical conceptions of men of the structure of the world and their position in it is a small history of philosophy. We show that early materialism was bound to fail because it possessed only one side of knowledge, even if it was very successful in this field. It could not explain much more than the naive classical dualism of body and soul. Karl Popper then proceeds somewhat more systematically, but still conservatively.

5.1 A Brief History of Philosophy

> "No man's knowledge here can go beyond his experience."
> John Locke, English philosopher of the Enlightenment, 1632–1704.

Physics tries to find the "Theory of Everything", the world formula or the theory of everything. By this is meant a physical-mathematical theory that summarizes and describes all elementary particles and thus captures all physical forces in one cast. This is not enough for a philosophical worldview: *We* belong to it, our psychology, our consciousness, our products such as art. All this is not physics. Let us briefly consider some approaches here ordered by the number of basic ideas in the world model.

- Monism:
  The classical materialistic attempt (end of the nineteenth century) of explanation with material processes. Here matter is the solid, naive concept of everyday life, represented for example by iron balls or wooden beams. The word "world formula" and the search

© Springer Fachmedien Wiesbaden GmbH, part of Springer Nature 2021
W. Hehl, *Chance in Physics, Computer Science and Philosophy*, Die blaue Stunde der Informatik, https://doi.org/10.1007/978-3-658-35112-0_5

for it originate from the already mentioned du Bois (1872) and are older than "the Theory of Everything".

The German zoologist Carl Vogt went down in history (of science as well as philosophy) with the shocking remark:

"Thought stands in the same relation to the brain as bile to the liver or urine to the kidney."

He had correctly stated as a zoologist that the mental faculties are necessarily bound to the brain. From then on his comparison becomes false: Thoughts are not material. But this he could only suspect, not know.

Chance is transparent to this phase of materialism: There are statistics for large numbers of random events, but one is sure that there is actually no such thing as chance.

- Dualism:

    The most outstanding recent exponent of mind-body dualism is the French philosopher René Descartes (1596–1650). He expresses what we all feel: Having a body and a nearly independent mind. This feeling brings with it the further sense (and illusion) that the mind exists independently of the body, even continuing to live as an "ME" after the body has passed away. His explanation is curious,

    There was simply no reason for her to perish with the body.

In the soul he sees a kind of substance whose main property is thinking:

"Que chaque substance a un attribut principal, et que celui de l'âme est la pensée".

"Every substance has its chief property, and for the soul it is thought."

He gets himself into a peculiar problem: Obviously body and mind work together - but how does that happen? He assigns this task to a gland, the epiphysis (pineal gland); it is logically suitable as a connecting element, since it sits in the brain and appears unpaired like a bridge. But of course it is a mistake.

The physical part of the world follows deterministic laws for Descartes, but the mental part is free for him.

- Trialism:

    As a modern, simple modeling of the world in three levels we consider the three-world theory of the Austrian-British philosopher Karl Popper (1902–1994). Karl Popper thus proposed a division of our world that extends dualism (Popper 1978). He calls it "the three worlds": World 1 is physics as the basis of everything, World 2 is the mental processes, conscious or unconscious. In addition to this classical dichotomy, he adds world 3: It is the world of culture, e.g. the content of libraries, mathematics, art. With Popper it is a rather arbitrary division (as he says):

"It's a metaphor that helps us see certain relations. You can't axiomatize such things; they are signposts, nothing more."

We will give meaning to the slightly different division.

A special feature is Popper's attitude to chance: he sees an effect through chance, which he calls "propensity" from the Latin *propense* 'inclined', 'tending to', i.e. an inherent inclination in chance. The idea looks esoteric at first, but we will see sense in it below.

Popper also leaves open, physically undefined, how the world works, deterministically or chaotically. His memorable question is:

"Are all clouds clocks or are all clocks clouds?"

This is the central question of our book!

According to its meaning, the monism of the ancient atomists belongs to this small history of philosophy as a curious precursor on the way to the "true" scientific world view. We have already introduced atomism quite extensively and empathically. Perhaps it is also more of a polyism because of the many different kinds of atoms! Whereas ancient atomism was pure (ingenious) speculation, nineteenth century monism or classical materialism, on the other hand, is a unique "evidence-based" success story, and atoms are in principle proven.

Already two centuries before, atoms had been more than a hunch. In 1648, the French physician Jean Magnen (Johann Magnenus) already estimated the size of atoms as very small, since the smoke of a grain of incense fills the space of an entire cathedral! However, with the first atomists of modern times (as well as the mathematician and astronomer Pierre Gassendi) chance disappears from the system; it is domesticated and Christianized and replaced by the will of the Creator.

The American naturalist and statesman Benjamin Franklin (1706–1790) had emptied a teaspoonful of oil onto the water on a pond in England, now a park in London - and estimated the area of oil produced to be 2000 m^2. The atoms existed and thus had to be really small.

The ancient atomists are at the beginning of the history of scientific chance. It is the first time that chance is relevant to the system! But for rationalists, such as the Roman politician Cicero (106–43 BC), the idea of chance is ridiculous (Goldstein 2007):

"What is this new cause in nature that makes the atom deviate? Or do they draw lots among themselves as to which shall leave its orbit and which shall not?"

The atomists would say something which the common mind must reject with contempt! However, it is analogously what Einstein will also say 2000 years later: God does not play dice.

But a very simple experiment says the opposite. It is the double-slit experiment: If you fire an electron at two slits that are close to each other, then it looks as if it goes through both slits at the same time, but then appears behind exactly one of the slits. The proof that both slits are involved is provided by repeating the experiment. Typical interference patterns gradually build up behind the slits: The particle is a wave. The experiment was carried out in 1807 by the English physicist and ophthalmologist Thomas Young with light. We are used to this with light, but the experiment also works with individual particles of matter: in 1959, the physicist Claus Jönsson demonstrated it accordingly with electrons. The double-slit experiment is considered one of the most important (and most beautiful) physics experiments ever.[1] It is central to understanding the role of chance in the world.

The above sentence by Cicero is exactly a poetic paraphrase of quantum physics: the particles decide among themselves through which of the two columns who will pass.

5.2   The Three Worlds of Karl Popper Updated with Chance and Software

5.2.1   Mechanical Machines cannot Think, Computers Can

"I propose to consider the question "Can machines think? . . . . Nevertheless I believe that at the end of the century the use of words and general educated opinion will have altered so much that one will be able to speak of machines thinking without expecting to be contradicted."
Alan Turing, British computer scientist, 1912-1954, in 1950.

Much has changed since Descartes (seventeenth century), since Turing (1950), and also since Popper (1978). Thus, in the first half of the twentieth century, relativity and quantum theory have changed many concepts, and these striking changes have penetrated the conceptual world of the educated, at least to some extent.

Other memes are less spectacular and more stubbornly resistant to change, such as the classical view of causality and chance not existing, and the understanding of the computer as a computational tool, and only a computational tool, that "only does what it (i.e. man) has been taught to do". The classical starting point of the human position towards the computer is:

---

[1] The experiment with electrons was voted "most beautiful" physics experiment by the readers of the magazine "Physics World" in 2002.

Humans are superior. "We" can do things XYZ that a computer never can. We don't know why, but only we can.

After all, we gave a (time-dependent) list of such questionable abilities above.

This position of supposed principled human superiority is chauvinistic and logically nonsensical: We are also a computer. Why should our technology in "flesh and bones" be better and inimitable? It is "human chauvinism", grown in pre-scientific times.

If we admit that we are a flesh and blood computer, we put ourselves in line with the animals, which generally have smaller computer models, and next to the digital computer. This explains why so many things are becoming interchangeable with the computer, including, perhaps to our regret, our jobs. "Human chauvinism" has grown up in the history of religions and philosophy and has also been understandable in view of our (previous?) uniqueness, but it is factually wrong in relation to intellectual and spiritual information technology. Animals have them too and, of course, increasingly better digital computers.

The above superiority thesis means technically-scientifically the assertion that man has an unnatural technology in the brain that gives him this superiority and which is scientifically inaccessible to us. This is pseudoscience. In the brain, there are many vague or precise program structures, hierarchies, fast-working, adapting and fast-learning parts, slow-acting, slow-learning and even inherited parts that altogether make up our identity. All these kinds of components exist in the digital computer as well!

The last bastion of human chauvinism, in today's discussions anyway, is consciousness: *"A computer can never have consciousness."* This is not only wrong in terms of information technology, but also philosophically.

In terms of information technology, our consciousness is first of all the operating system of our "computer system", which enables us to live in our world. This operating system is for the most part unconscious. For this we have internally the possibility to observe and linguistically follow the work or at least the call of some of our inner software routines. An example of this is the reading of reading material, here the process of reading the word "consciousness" or "consciousness":

- In the beginning, there are signals "black and white" from the retina, in the digital called pixels, actually first in color, from which is immediately abstracted.
- This gives rise to structures such as straight lines with loops that identify letters that combine to form words.
- Then you see a word with "ss" instead of "ß" and you may think, "The author is Swiss?"
- Now you are in the conscious part and thinking ahead.

This is all pure, conventional information processing. Your digital computer can do this too and does it very similarly, just in different technology.

The American philosopher Daniel Dennett (born 1942) puts it succinctly:

The mind is the effect, not the cause.

The mind is a kind of running software, no more. The naive notion of the mind as an independent power that acts directly led to ideas such as telekinesis (the movement of another body by thought alone), telepathy (the transmission of information from or to another by thought), the survival of the mind after death, and spiritualism, which brings all these concepts together.

Famous (as an error) are the experiments of the physicist and parapsychologist Robert Jahn (1930–2017). Jahn wanted to influence chance by thinking. He had built an electronic random number generator for this purpose. The test subject looked at the number display of the running device and had to think: "Higher" (in English thinking probably). Jahn was sure it would work, the numbers generated would go higher. He failed because of the pitfalls of experimenting with randomness. There is no influence, his work has gone down in the history of pseudoscience.

Many people have a reluctance to accept that we and they themselves are a "computer": After all, we are much better, and besides, somehow special, somehow supernatural. But we are not better, we are special, but not supernatural. Our body and our mind follow the laws of nature including those of IT. We are just a completely different implementation, i.e. a different design for the same basic purpose. I urge to have the courage and think of ourselves that way. It makes many things easier.

We emphasize that this is not "physicalism", that is, an explanation out of physics: physics and information are, as stated in the introduction, two different fundamental worlds. The Microsoft Word program used to write this text has nothing to do with the transistors of the labtop computer (except to use them). On the contrary, for all living things, information, the bit-world, is more important than the It-world of physics. A possible neutral name for this explanation of the world is *naturalism,* with its emphasis on the rejection of the extra-natural.

Consciousness is a clear, conceptually simple IT process. The peculiarity is philosophical: The I does not stand *next to* the brain and hear it say, think and decide, but:

The brain *is* the ego.

The philosophical fallacy that consciousness is something inexplicable is called the "qualia problem"as a philosophical question. The word qualia from the Latin *qualis* 'how is constituted' means here the subjective sensation of a property, e.g. one of the colours in Fig. 5.1a, for instance the blue of the single blue pencil. The term *qualia* sounds scholastic, but it first appeared in materialism, in philosophy in 1866 with the philosopher Charles Peirce - indeed, you have to have a Descartesian dualist worldview to see it as a problem at all. Gottfried Leibniz describes it so beautifully mechanistically: If the brain were as big as a mill, into which one could enter,

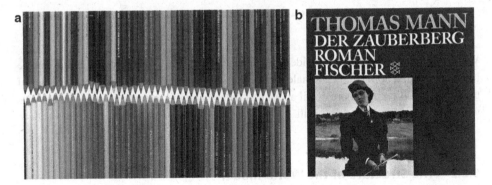

Fig. 5.1 Two examples of objects for mental impressions, "elementary" or "higher". Image (**a**) Color impressions. (Source: Crayons, Wikimedia Commons, KLJ). Image (**b**) A classic book. (Source: eigen/Fischer)

Then one will find nothing but pieces that push against each other, but never anything from which one could explain a perception.

The brain is not mechanics, but the sensors are psychophysics and electrochemistry, the brain electrochemistry and especially computer science. This sounds as bad as "the brain secretes thoughts like the kidneys secrete urine", but only on the surface. It's a fundamentally different world, the world of computing in the generalized sense. Without understanding this, no mind, however clever, can cross the chasm from mechanics to intellect to soul, not Descartes, not Leibniz, and not yet Popper. My sensation of "blue" goes through *my* sensory system, *my* processing in the brain with *my* software, *my* memories, and based on that to *my* associations. Why should your, yours, his chain of processing be the same? It is different, of course.

This becomes even clearer when looking at (Fig. 5.1b), a book cover of Thomas Mann's novel. Each reader has his own chain from seeing the letters to the associations of the sanatorium in the Swiss Alps. The American philosopher Daniel Dennett (b. 1942) has rightly called qualia

"an unusual expression for something that couldn't be more commonplace for all of us: How things appear to us."

One might add "an unnecessary but distinguished expression." Dennett, the great philosopher, considers qualia to be an example of the tendency of philosophers (the "philosopher's syndrome") *to make an insight into* a *necessity* out *of a "lack of imagination."*

The fundamental philosophical problem of Descartesian dualists (and many people still are) with the computer is that they see the computer as a mechanical aid to the ego, which stands as an aid on the non-mental, stupid side next to the ego. But the I is not standing *next to* the brain and listening to it say, think, and make suggestions, but:

The brain is the I. I, the brain, think and decide. This is simply unthinkable for many people.

We are a computer from the inside! The idea that the ego is next to the brain can be illustrated with a homunculus, a male that in truth makes our decisions in a back corner of the brain. We'll go into more detail about this fallacy and the metaphor of the homunculus when we discuss so-called free will.

5.2.2   The Structure of the World Model

"I will propose a view of the universe that recognises at least three different but interacting subuniverses."
Karl Popper in the 1978 Tanner Lectures.

We intend to update the world model on Popper's basic idea. Especially the new understanding of information technology as a basic technology also of life is a driving force.

Like Karl Popper, we define physics and its derivatives, such as astronomy, chemistry and geology, as the first pillar of the world. This remark is not arrogant, but means, for example, for chemistry: Chemistry is the branch of physics that describes the properties and transformations of various combinations of atoms and their electron shells in the range of a few eV (electron volts).

The first pillar is world 1 with all phenomena of the inanimate world, which develop out of themselves like snowflakes out of cold, damp air. Then we define as the second pillar or world 2 everything that is built on an explicitly stored blueprint that can be executed. The pillar is based on nature's inventions of passing on something successful and not having to start from scratch every time. Of course, this requires a device that can read and process this knowledge. Such a device is, in a general sense, a computer. The second pillar includes everything that is "computer-like" or produced by a computer-like device.

- The plans that are processed are, so to speak, the software of the computer.
- The unit that exploits and implements the knowledge is the hardware.
- The rules that connect the two is the "architecture".
- The working off itself is the spark of life that goes on from command to command.
- In addition, there is chance - more so in nature, less so in the digital computer.

Physically speaking, the "computer" runs apart from the thermodynamic equilibrium, which occurs only at death.

This definition includes all life and all products of digital information technology. Since we include consciousness and soul, our world 2' overlaps here with Popper's world 2. But we also position here what was formerly mystical and for Popper the spiritual world 3, such as language, knowledge and the products of language such as speech recognition, understanding and translations. There is a simple and objective criterion for what is the object or process of world 2′:

Everything that a digital computer can master in principle is also world 2′.

An aside on this: It is an extraordinary achievement of the ancient atomists that they had already clearly identified a world 2 separate from physics, i.e. from substance. Perhaps the finest example of this comes from the late atomist Galileo Galilei as *"the tickling of Galilei"* from the year in the 1623 writing "Il Saggiatore" (Hehl 2017):

Galilei shows in this the difference if we tickle the foot of a human being with a feather or the foot of a marble statue: The tickling is obviously not a property ('substance') of the world 1 objects 'feather' or 'marble foot', but arises in the human being. The tickling is not a substance and the human being is something else, they are world 2′ objects.

This distinction between World 1 and World 2′ can also be found to some extent in ancient atomistics and its premonitions. To the two worlds correspond different types of atoms, to world 1 coarse, jagged and heavy atoms, to world 2′ fine, smooth, ideally spherical and easily volatile soul atoms. In today's language, this fits the distinction between 'it-world' and 'bit-world'. The soul atoms are information.

This substantially limits the scope of Popper's World 3, and even calls it into question as a whole. We see this in the example of art, which we re-evaluate vis-à-vis Karl Popper. Since about 1965 computers have been composing music and creating pictures - with the help of chance. It is no longer at all easy to distinguish a poem or a watercolor by a computer from human products! With 3D printers, computers even create sculptures that would be impossible for a human. Art is, at least to a good approximation, an information product of rules, memories and chance.

We use the same example as Popper, Beethoven's Fifth Symphony (Fig. 5.2), to illustrate all three worlds. The first step, composition, is an act of creating a World 2′ object from rules, memories, moods, and chance. The resulting score is world 2′, written and

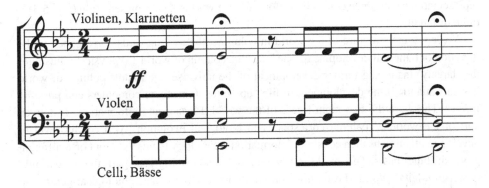

Fig. 5.2 The opening motif of Ludwig van Beethoven's Fifth Symphony. Image: German Wikipedia, Fifth Symphony

printed as a world 1 object. The creation of the sounds is physics and therefore world 1. The conductor directs the orchestra according to rules, the score, according to his understanding - it is world 2′ with chance.

The identification of world 3′, on the other hand, is critical: Do *you* say *"it was quite beautiful"* or *"a miserable performance"* or do you say *"a great work of art"*, *"wonderful as if from another world?"* In the latter case we give the work (the score, the performance) the mystical character of world 3′.

So there will be people who see Thomas Mann's "Magic Mountain", van Gogh's "Starry Night" or Beethoven's "Fifth" as a World 3′ object - but for others the Magic Mountain is a report, the Starry Night a coarse painting and the symphony just long-winded noise. The same applies to the possible World 3′ object "love": just hormones, self-interest and posturing, or a superhuman force. World 3′ is therefore bracketed.

Everything constructed is thus computer science and world 2′. A useful subdivision concerns our knowledge of the inherent blueprints. In one part, let us call it world' 2 a), the blueprints are accessible, for instance in biology or in many programs in the digital computer. In world 2′ b), the running software is not immediately apparent, for example in our psyche, but can only be divined by experiment or by physical methods such as the imaging techniques of neuroimaging.

A big difference of the "computer brain" to the usual digital computer concerns the coincidence and its role. In the digital computer, physical randomness (i.e., the error) should not appear in the single run, unless it is needed and artificially generated as a pseudo-randomness. Given the millions or billions of instructions the computer processes per second, even very rare errors would continuously crash the computer. It is (actually) the goal of any software development to deliver a product that is as error-free as possible as quickly as possible.

It's different in the human brain; here chance plays a big role. Therefore, we will take a closer look at two areas of human IT below: Decision-making and "free will" on the one hand, and creativity on the other.

If you google for pictures of Karl Popper's Three Worlds Doctrine, you will usually find representations with three equal circles for the three worlds. Our representation (Fig. 5.3) is quite different:

We depict the structure of the world in columns; the axis upwards corresponds to the direction of temporal development. The left, blue column of world 1 ("physics") illustrates the (largely) inanimate further development of the universe, the middle column of world 2′ ("information") the development of life "up" to us humans, to computers and possibly further. These are the two main forces of the world: Physics and information. The green part of the world 2′ column corresponds to biological evolution. At a certain level of development, humans then construct information technology themselves, but on a different physical (blue) basis. This marks the beginning of the leaning grey part, the increasingly powerful digital world. From now on, both world 2′ variants, biological-green and technical-grey, run side by side and closely together further with an uncertain future.

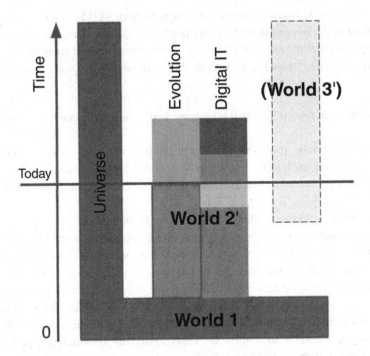

Fig. 5.3 The updated three-world model. World 1 comprises the inanimate (physics), world 2' everything constructed (IT or software in the extended sense) and the optional world 3' the "higher values" such as art. Chance, ultimately coming from world 1, helps determine development in all worlds. Blue: physical basis Green: biological IT Grey: technical IT

One possibility of thought is the fusion of biology and technology as "transhumanism", in the worst case with the disappearance of biology.

All objects of world 2' need a (blue) foundation in world 1, for example in the form of proteins in biology, neurons or transistors in the case of the digital computer, and an associated world 1 energy source. Therefore, any world 2' is positioned on the foundation of world 1. This represents a big change in philosophy. Mind is not a substance and is not independent.

> **Thus two continuously growing worlds emerge side by side. For us as humans, the world of the constructed, of information, is essential, built on top of the world of physics. The world of the constructed (of life) is a continuum, starting from chemistry "up" to us and to digital computers. The continuum is open upwards.**

Chance, starting from world 1, pervades all three worlds. He even builds world 2' himself, perhaps also the products of world 3' like art! The interface between world 1 and world 2' of the world model, green on blue, is critical. This is where the idea of spontaneous generation was localized ("life just comes into being"); today it is the problem of chemical evolution (see below). There are important feedbacks from world 2' into world 1, such as the generation of the oxygen atmosphere by plants and climate change by humans.

We emphasize that physics and information have a fundamental connection that is not understood. This is expressed, for example, in the fundamental question (Aguirre et al. 2015):

"It from bit or bit from it?" - What is primary, matter or information?

This is perhaps the most profound question in physics. For many essential questions of philosophy, however, such as the body-mind problem, it has little meaning: Our mind is a function of the complex software structures "high above", not of the physical foundations "deep below".

The new three-world-model allows a clarification to the schools of thought of the great ancient Greek philosophers Aristotle and Plato, already mentioned several times. Plato sees the world from the point of view of "ideas", the idealized models to things. Aristotle builds the world from "below" and attempts to explain the world from experience using a precursor form of the scientific method. Thus the two philosophers have extreme positions of thought in the world model and different schools of thought: Plato thinks from the top down, Aristotle from the bottom up. In the computer analogue, Plato thinks primarily in terms of the specifications of objects as "ideas", Aristotle systematically examines the foundations and attempts to build from the bottom up, starting with physics, thus creating the foundations of science.

World 1 is also capable of building very complex structures without an external blueprint, such as the world of stars, planets and galaxies. This results in the typical L-shaped structure in the graphic. The crossbar of the "L" of physics carries the IT world.

We assume that possible world 3' objects are based on our thoughts and experiences, i.e. on a sufficiently sophisticated world 2'.

From now on we will mainly talk about world 2' and world 3', not about Popper's primal forms. We will, however, keep the dash (') to let the difference remain clear.

5.2.3   Chance Is Necessary

"An interesting variant on the idea of a digital computer is 'a digital computer with random element'."
Alan Turing, British mathematician, 1912–1954.

Alan Turing was aware of the importance of randomness for mathematics: True random numbers cannot be calculated by algorithms. He had therefore succeeded in getting an electronic random number generator built into the Mark I computer, but it obviously did not work well. His desire to incorporate the random number generator probably seemed to his colleagues to be just as artificial as the incorporation of randomness as a clinamen, a constant trembling, into Epicurus' atomic world 2200 years earlier!

Turing has a mathematical and physical interest in chance. He sees "real" chance as something beyond computability about which he has many thoughts. Thus, randomness is related to his little-known philosophical idea of a magical "hypercomputer" that he calls oracle (Prisco 2018; Hodges 2019).

▶ **Definition   The oracle of Alan Turing is a black box, which delivers mathematically transcendent things, e.g. non-computable numbers or real coincidence.**

There is for Turing a kind of transcendence in mathematics associated with the names of Kurt Gödel and Gregory Chaitin. Real chance and non-computable numbers are part of it. In the number continuum, the vast majority of numbers are non-computable, i.e. there is no rule to compute them! In physics, in this sense, all sources of randomness (the sea, the noise of electronics, the initial fluctuations in the Big Bang, or Darwin's pool (see below)) are such oracles.

Coincidence probably also interests him for another reason. Alan Turing believes in parapsychological phenomena, especially telepathy. It is the heyday of (futile) parapsychological research, for example in the famous Institute for Extrasensory Perception of the American botanist Joseph Rhine from 1931 on. A typical experimental technique was, for example, the guessing of hidden playing cards by human mediums. The idea is to see if a medium guesses better or "somehow" sees which card values will be drawn next than chance predicts. Science will never be able to verify and never accept the parapsychological results claimed by Joseph Rhine. But Alan Turing thinks telepathy is proven. The random number generator would have been a preparation for such experiments.

But there has already been another arbitrary incorporation of chance into a theoretical edifice, around the beginning of the nineteenth century. It is a marginal note in the history of science: In 1900, the German physicist Max Planck (1858–1947) had empirically found the formula with which intensity a body emits thermal radiation of different colours. But there was a problem in deriving this formula in classical physics (Grimsehl 1988). The variety of oscillators for thermal radiation was in each case independent of each other and remained so, since all equations were linear. Figure 5.4a symbolizes this with a mirror box with aligned light rays representing ordered standing waves.

Planck introduced a trick to achieve coupling: A small dust of charcoal is placed in the box (Fig. 5.4b). The small grain is a microscopic oracle in the above sense. It transforms order into disorder or chance. It absorbs and emits light of all colors chaotically. Thus the radiation is distributed and the disorder of thermal equilibrium results. Planck was able to derive exactly the desired formula.

Whether Democritus, Planck or Turing, without chance the world is boring, because it is ordered. Without chance the world would be unnatural, not nature.

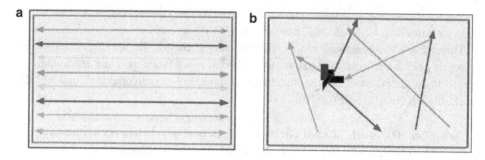

Fig. 5.4 Planck's carbon dust brings randomness. (**a**) An ideal classical box without randomness. The light paths are well ordered. There is no thermal equilibrium. (**b**) A small, irregular coal dust brings the coincidence into the box. This results in thermal equilibrium

Chance does not initially appear in the diagram (Fig. 5.3), not even in Karl Popper's: It all looks like orderly growth. This is completely wrong. For this we imagine the picture of the world without chance, totally ordered:

- All the leaves on a tree would be the same, indeed all the leaves of a species,
- large groups of people would be the same,
- the walls would be smooth as glass,
- the waves on the lake or sea would be well-ordered and in step,
- the clouds would be clean spheres,
- the stars are evenly distributed in the galaxy or lined up in the spiral arms, and so on.

Of course, without chance, strictly speaking, there would have been no species, trees or humans, because without chance there would be no evolution. The picture of the world without chance is absurd. We stand here like the ancient atomists who introduced chance to destroy the triviality of order and to make the world alive.

Chance can intervene in the course of the world in many different ways. Figure 5.5 illustrates a "particularly coincidental" possibility. The interstellar object Oumuamua was proven to have come from outside our solar system! If it had hit the earth, this would have been the perfect coincidence in world 1, which would also have been a tragedy for world 2′!

Chance hits the world 2′ in two ways: Out of the physical foundation or within the information world itself.

A good physical example are the cosmic rays, i.e. about 100,000 electrically charged high-energy particles per square meter and second, which reach the earth's surface. Since about 1980, the electronic components of computers have become so small and thus so sensitive to interference that individual bits in computer memory can be "knocked over" by such particles of cosmic rays. This is largely prevented by built-in automatic corrections. The cosmic particles as well as the otherwise existing natural radioactivity also penetrate our bodies and damage our IT, cause mutations and cancer.

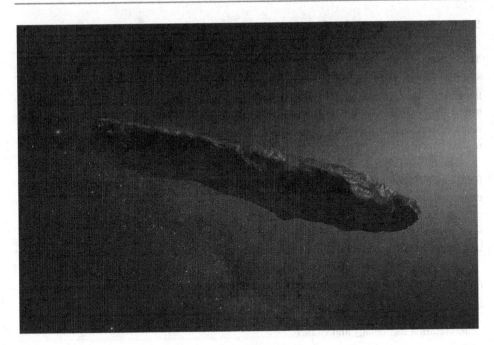

Fig. 5.5 The possible absolute coincidence: A celestial body comes from far out in space. Artist's impression of Oumuamua, the first known interstellar object. Image: Artist's impression of Oumuamua, Wikimedia Commons, ESO/M. Kornmesser

Coincidences at the IT level within World $2'$ are manifold: when the gametes of two people who met by chance unite, it is a coincidence that can change the world.

In the technical IT world, it can be a misrecognition and misinterpretation of data, a random result of a Google search on an input that has changed only slightly, a blocking by another process that happens to be running, and much more. The Google search algorithm finds millions of hits on many queries, picks out thousands of them, flushes a dozen onto the user's first screen looking at the top three suggestions: The Internet is a giant informational random machine with slants, explicitly guided with advertising.

In particular, the randomness in world $2'$ may also be intentional randomness from a physical randomness generator or a mathematical pseudorandomness generator. We discuss such methods below.

Even on the assumed level of world 3, an effect of chance is conceivable: For example, when Beethoven's digressive thoughts and emerging memories change the flow of the composition during the composing process.

We find the greatest effect of chance at sensitive decision points both in inanimate nature and in the human brain, but especially in the concentrated information in the genetic material in biology. There, an elementary atomistic random act can cause a dangerous change and thus perhaps become the starting point of a pandemic.

5.2.4   Summary of the Chapter

The historical conceptions of men of the structure of the world and their position in it is a small history of philosophy. We show that early materialism was bound to fail because it possessed only *one* side of knowledge, even if it was very successful in this field. It could not explain much more than the naive classical dualism of body and soul. Then, somewhat more systematically, but still conservatively, Karl Popper proceeds, distinguishing three realms (worlds). It is the world of things, of feelings and cultural goods.

A gem is the oldest of the philosophical models we consider, ancient atomism. It is a foreshadowing of modern times with matter and transformation, soul and even built-in chance. But no philosopher had a chance to create a correct and meaningful model until the computer was recognized in its philosophical significance. Until then, soul and intelligence were mystical concepts. Deeper understanding of this side of the world really only begins with Alan Turing (1950).

The author asks us humans not to be shy and to accept that we are also computers, but realized quite differently. We are nothing supernatural, but still something special.

We present a trialist model of the world, similar to Karl Popper's, now updated with information technology and chance:

Physics for the inanimate, IT (computer-like) for everything animate and constructed, and if necessary as a third something superhuman-cultural like art and love - if you believe in "real" art and "real" love.

Something essential comes in addition and is necessary for the construction of the world: There is coincidence in the physical world and the IT world. There are small jumps in the course of events everywhere, just as Epicurus already attributed to the atoms. Just try to imagine a world without randomness! It would be a crazy world, with identical leaves on the trees, orderly waves, and perhaps exact spherical clouds. Of course, there would be no species and no human being.

In mathematics, non-computable numbers and non-decidable assertions correspond to chance. The computer scientist Alan Turing introduced the concept into the philosophy of mathematics as an oracle.

Chance is built into the foundation of the world.

References

Aguirre, Anthony, et al., eds. 2015. *It from bit or bit from it? On physics and information.* Heidelberg: Springer.

Goldstein, Jürgen. 2007. *Kontingenz und Rationalität bei Descartes.* Hamburg: Meiner.

Grimsehl, Ernst. 1988. *Lehrbuch der Physik, Band III, Optik.* Leipzig: Teubner.

Hehl, Walter. 2017. *Galileo Galilei kontrovers.* Berlin/Heidelberg: Springer.

Hodges, Andrew. 2019. Alan Turing. *The uncomputable*. The Stanford encyclopedia of philosophy. plato.stanford.edu/archives/win2019/entries/turing. Zugegriffen im Juni 2020.

Popper, Karl. 1978. *Three worlds: The Tanner lecture on human*. Values.tannerlectures.utah.edu/lecture-library-php.

Prisco, Giulio. 2018. Karl. 1978. The Turing oracle. *Creative randomness from the beyond*. turingchurch.net.

Turing, Alan. 1950. Computing machinery and intelligence. *Mind* 49: 433–460.

Evolution: The Creativity of Nature

6

Teilhard de Chardin is (was) an extraordinary personality, Jesuit, scientist and Christian philosopher (Fig. 6.1). From the point of view of the official church and the superiors of his order he was a lateral thinker, because he tried to integrate and map the whole evolution into Christianity as the way to an ideal state, which he called the omega point. Presumably some of his mystical statements are appropriate for discussion as potential "world 3'" objects, such as "love" and "the feminine." He even writes a little book *L'éternel féminin* about the eternal feminine, quite similar in theme to the author's (Hehl 2020) about the women in his life.

6.1 Evolution Is Not a Theory

6.1.1 Teilhard de Chardin

> "Is evolution a theory, a system, or an hypothesis? It is much more: it is a general condition to which all theories, all hypotheses, all systems must bow and which they must satisfy henceforward if they are to be thinkable and true.
> A light that illuminates all facts, a curve that all lines must follow, that is evolution."
> Pierre Teilhard de Chardin, French theologian and philosopher, 1939.

The author of the long quotation is (was) an extraordinary personality, Jesuit, scientist and Christian philosopher (Fig. 6.1). From the point of view of the official church and the superiors of his order he was a lateral thinker, because he tried to integrate the whole evolution into Christianity and to map it as the way to an ideal state, which he called omega point. Presumably some of his mystical statements are appropriate for discussion as potential "world 3" objects, such as "love" and "the feminine." He even writes a little

© Springer Fachmedien Wiesbaden GmbH, part of Springer Nature 2021
W. Hehl, *Chance in Physics, Computer Science and Philosophy*, Die blaue Stunde
der Informatik, https://doi.org/10.1007/978-3-658-35112-0_6

Fig. 6.1 The theologian and anthropologist Pierre Teilhard de Chardin in a cave in Castillo, 1913. Image: German Wikipedia, Claude Cuénot

book *L'éternel féminin* about the eternal feminine, quite similar in theme to the author's (Hehl 2020) about the women in his life.

We agree with the content of the above quotation; we have already mentioned the short version by Theodosius Dobzhansky:

Nothing in biology makes sense *except in the light of evolution.*

The often heard word "evolution theory" is chosen rather unfortunate, because in common usage *"theory"* is understood as an unconfirmed thesis. The concept of evolution has this fate in common with the theory of relativity, both special and general. Also in these terms a doubt resonates, but this is not appropriate for evolution as well as for the theory of relativity, both - evolution in biology and relativity in physics - are the most important and most certain areas of the respective science!

6.1.2   Young-Earth Creationism

"Creationists make the word 'theory' sound like something dreamt up after being drunk all night."
Isaac Asimov, Russian-American scientist and author, 1920–1992.

Evolution owes its controversial reputation, at least in the nineteenth century, to the competition of the literal and naive interpretation of the biblical story. According to it, all plants were "made" on the third day of creation, all aquatic animals on the fifth day, and all land animals (and man) on the sixth day.

In addition, there was no lack of attempts to determine the corresponding year with biblical-historical analysis. The most famous date comes from the Irish Archbishop and

Primate of Ireland James Ussher (1551–1656): According to him the creation began on Sunday, October 23, 4004 B.C. Another important pseudo-date is the day of the landing of Noah's ark on Mount Ararat, it "was" Wednesday, May 5, 2348 B.C. According to the claim of the Bible Noah saved all animals in pairs. In Ussher's time, the attribute "all" was not seen as an insurmountable problem, but with the discovery of new, exotic lands, the number of known species increased dramatically. The question of the animal and plant occupants of the ark passed into the scientific study of the global animal and plant world, into natural history. Another problem was the need to re-colonize the world according to the biblical story. All animals were local to one place on Mount Ararat. One possibility discussed was that after the fall of the Tower of Babel (according to Bishop Ussher 106 years later) the scattering peoples took their animals with them. Even the English philosopher and polymath Thomas Browne (1605–1682) wondered in 1646:

"Why then did the natives of North America take the rattlesnakes and not horses? Very strange."

So the creationist theory (outdated to the enlightened theologian) says:

- The creation occurred in one fell swoop, the timing was, geologically speaking, recent,
- all animals and plants could be concentrated in a very small space,
- the number of animal species is typically 100, the number of plants in the Bible is about 40,
- the plant and animal species are considered closed and constant.

To the biblical animals even exists a Wikipedia article "List of Animals in the Bible". And another important point of the creationists for us humans and for the animals, unfortunately often to the disadvantage of the animals:

- Man is not an animal, but something quite different.

Overall, this creationism may have been humanly satisfying, but from today's perspective, it is childish thinking. There has hardly ever been a more nonsensical "theory".

Table 6.1 shows today's estimates for the numbers of animal and plant species. The numbers have now become serious and exceed the naive understanding of an ancient nature lover by orders of magnitude. The same applies to the other points: The evolution took place over four billion years, took an astronomical number of acts of creation, was scattered over the earth and earth history, the number of species was and is many millions of animals and plants, and the evolution was dynamic and continues today.

The working space of evolution is these millions of life forms and the associated many trillions of individual living things. This is to be compared with the smallness of the idea of biblical creation, if taken scientifically.

A witty classical argument against evolution and for a finished creation comes from the English theologian William Payley (1743–1805). It argues against the workings of chance

Table 6.1 Numbers of known species on Earth

7.77 million 953,000 thereof	Animals, described and catalogued
298,000 215,000 thereof	Plants, Described and catalogued
611,000 43,000 thereof	Mushrooms Described and catalogued
36,000 8000 thereof	Protozoa (motile protozoa), Described and catalogued
27,500 13,000 thereof	Chromista (plant protozoa) Described and catalogued
8.74 million	**Eukaryotes (multicellular organisms with cell nucleus)**

Data according to ScienceDaily, August 8, 2011

and for a finished, divine blueprint of everything. The argument appears in the latter's 1802 book "Natürliche Theologie". Darwin knew the book, and thus the analogy, well. He told a friend in 1859, *"I used to be able to recite it almost by heart."* Paley's parable has gone down in the history of philosophy as the "watchmaker analogy" (Fig. 6.2). The nonfiction bestseller "The Blind Watchmaker" by British biologist Richard Dawkins (b. 1941) wittily alludes to the watch paradox.

The crux of the parable is the ontological[1] difference between a stone and a clock, both lying on the path through the heath. The stone is a part of nature, always there or at least long there, the watch obviously the product of an intelligent man, a watchmaker, and not from the world of the heath but from civilization. Paley concludes that anything sufficiently complex needs a designer and a plan.

The designer has the meaning of the device in mind, sketches the blueprint, and makes the watch according to it. Paley and most of his contemporaries consider this a teleological proof of God. For us it is a pretty example of the three worlds doctrine and a glimpse ahead into the meaning of evolution:

Evolution has built this dangerous thought *"is there another human (or animal) at work here that I don't see"* into our psychology as the "agent detection effect". It would have been dangerous to miss such a sign. A bit more about the evolutionary heritage in our psychology in the chapter "free will".

The stone is inanimate, a product of physics, chemistry, geology, etc., and thus a (random) object of world 1. The watch is based on physics with a base of steel and gold and is a constructed object of world 2′, similar to other products, such as a software program with a given purpose (the specification of the product), a development and realization and finally the application (the program run). But of course the plants, animals and we humans are living beings with a blueprint, i.e. world 2′. Paley also still considered the animals as a kind of machine, but of course not us humans.

---

[1] Ontological from ancient Greek ὄν *ón* 'being' meaning 'in essence, according to essence'.

a   CHAPTER I.

STATE OF THE ARGUMENT.

In croffing a heath, fuppofe I pitched my foot againft a *ftone*, and were afked how the ftone came to be there, I might poffibly anfwer, that, for any thing I knew to the contrary, it had lain there for ever: nor would it perhaps be very eafy to fhew the ab-furdity of this anfwer.   But fuppofe I had found a *watch* upon the ground, and it fhould be enquired how the watch happened to be in that place, I fhould hardly think of the anfwer which I had before given, that, for any thing I knew, the watch might have

VOL. I.   B   always

WORLD 1 – object   WORLD 2' – object

Fig. 6.2 The "watchmaker's paradox" of the theologian William Paley. (**a**) The original text (excerpt). Image: Paley Natural Theology, Wikimedia Commons, google books. (**b**) An object of the world 1: a stone. Image: own. (**c**) An object of the world 2′: A clock. Image: Montre Gousset, Wikimedia Commons, Isabelle Grosjean

The watchmaker could make the watch after maybe three years of teaching, but evolution developed "us" and our blueprint in four billion years and gave us (more or less) sense *a posteriori*. Plan and sense evolved together for the whole as for the individual functional components. It is the solution to the question that Paley does not ask himself: Who made the intelligent designer for the living world? Evolution evolves complexity itself, out of itself. But all of this and all of the associated fundamentals (age of the earth, genetics and proteomics, information technology, etc.) were still unknown and unimaginable. It is all the more magnificent that Darwin, without having all this at his disposal, was able to make a scientifically impeccable start.

6.2   Evolution as Software Technology and Process with Chance

6.2.1   The Principle

"Evolution is an up-hill random walk in software space."
Gregory Chaitin, Argentine-American mathematician, born 1947.

Technically, an evolution in the narrow sense is the run of a software, which uses a random technique with systematic experiments to reach a goal without deeper understanding what it does. In the big, biological evolution there is no visible goal, but only the minimal requirement "continue in the system, almost no matter how".

With this we need for the start of an evolution:

- an optimization task,
- a procedure for the systematic generation of candidates,
- a way of assessing the quality of a candidate
- a strategy for moving on.

A simple method for moving forward is to use the points in the immediate vicinity as the next candidates in each case and examine them for improvement.

Take as an example the small antenna in Fig. 6.3. This antenna was developed with an evolutionary algorithm for the program "Space Technology 5" as a *"very small, very*

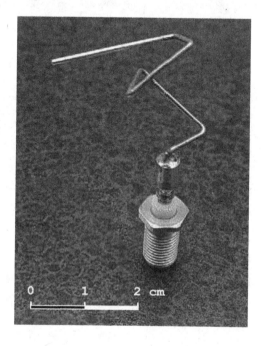

Fig. 6.3 An evolutionary
developed product. Image: ST
5x Band Antenna, Wikimedia
Commons, NASA

improbable looking, but very promising antenna", as its product description says. The task was to optimize radiation and reception characteristics with minimal weight. Testing the goodness of a proposed antenna can be done experimentally in a computer or in a model. The result was a new design, the likes of which no engineer had thought up before: chance created it in a controlled way, chance is creative.

With this criterion for the candidate, we get a figure of merit or a measure of the fitness of the candidate in terms of our task, and based on that we generate the next candidates and repeat the test until we are satisfied with the merit or a resource is used up, say time runs out. Evolution in this simple case is a partially-directed odyssey through an unknown mountain range. The success of a step is always confirmed only in retrospect. The end result is something new, an emergence. The term comes from the Latin *emergere* - 'to emerge, to come up' and means in more recent philosophy that a higher level of being emerges from a lower level. Evolution can thus do something incredible: it creates something new by chance.

For simple tasks, one could think of checking all possible candidates together and thus have the absolute best solution; this would be solving with "brute force", with brute force of calculation. With a handful of criteria that encompass all possibilities for a solution, one can even do this on a sheet of paper: It is the procedure of the "morphological box" of the Swiss astrophysicist Fritz Zwicky (1898–1974) as a method for solving a problem.

To do this, one creates a matrix of the features that are important for the problem on one side and the associated attributes on the other, and analyzes all combinations. Experience shows that most of the combinations will be meaningless, but one or two might work in a way that had not been thought of before. This process is considered a "creativity technique." But it is visibly rigid and thus not actually creative.

An even simpler (and much older) variant, on the other hand, seems more creative, although it is pure mechanics, but with chance: the mechanical construction to find new candidates. It is the Ars Magna, the great art, of the Mallorcan theologian and philosopher Raymond Llull (1232–1316). The goal is to gain divine wisdom through recombination. Seven rotatable concentric rings with terms are arranged on a disc (Fig. 6.4). As one rotates, one generates a new set by chance. The "figure of merit" is the degree of the "spiritual impression" of the new sentence.

But many problems have too many possibilities and internal degrees of freedom. The search space is also much too large for the computer to examine all points.

6.2.2 The Raw Coincidence

This highlights the well-known "problem of the endless typing monkey" (Fig. 6.5):

A monkey sits in front of a typewriter (or several monkeys in front of several machines, it makes no difference) and writes or presses random keys. Can this, for example, randomly produce Shakespeare's Hamlet? Or, if he could only write digits, randomly obtain the first 15,000 digits of the number π, say 3.1415926, etc.? For the latter task, there would be a

Fig. 6.4 A mechanical process for generating new by recombination. Turntables by Ramon Llull, Ars Magna, thirteenth century. Image: Ramon Llull, Fig. 1, Wikimedia Commons

Fig. 6.5 The typing monkey as an example of raw chance. Early Office Museum and New York Zoological Society. Image: Monkey typing, Wikimedia Commons, New York Zoological Society

completely different possibility for a random solution: could the monkey randomly write a computer program that would then calculate the number π?

The classic formulation of the memorable image of the writing ape is said to have occurred in Darwin opponent Bishop Samuel Wilberforce's famous June 1860 argument with Darwinist and biologist Thomas Huxley. It is the conversation in which the bishop is said to have asked the famous sarcastic question, *"Are you actually descended from the ape by way of a grandfather or grandmother?"* This question, and Huxley's analogous reply, *"I would rather be descended from the monkey than from them,"* are vouched for.

Huxley is said to have recklessly argued in favor of blind chance:

> **"Six eternal apes, randomly striking the keys of six eternal typewriters with unlimited amounts of paper and ink would be able to produce Shakespearean sonnets, complete books, and the 23rd Psalm."**
> **German Wikipedia, Infinite monkey theorem, pulled June 2020.**

This is not a trivial argument, probably not a "real" quote either, although there was a patent on a "typographer" as early as 1829!

The monkey (so intelligent in itself) (Fig. 6.5) here stands for stupid, undirected chance. It is an immensely impressive idea and an interesting intellectual problem!

It is even found in Cicero as an argument against the absolute chance which he imputes to the atomists:

> **"[An atomist would also believe] if innumerable specimens of the twenty-one letters of gold [. . .] were poured out on the ground, the Annals of Ennius would come into being, all ready for the reader. I do not know whether chance would succeed in producing even a single verse."**
> **Marcus Tullius Cicero, Roman statesman and politician, in "de natura deorum", 45 BC.**

The strict answer is yes, it could be, but it's not a good argument for evolution. Cicero is effectively right here. The affirmative gives a completely false sense of probability. The unlikelihood in these crude examples of creating something meaningful by blind chance is incredible. The Argentine writer Jorge Luis Borges (1899–1986) is fascinated by this quasi-infinity of random letters and goes further. He uses it to invent a magical library of total randomness. This *library of Babel* would contain all the 410-page books possible by letter combination, and thus all the knowledge in the world, though among them so much nonsense that *"all the generations of mankind could have gone through the shelves before only one tolerable page was found."*

If Huxley should have really said the above assertion, then the point would go to Bishop Wilberforce, even if Darwin's teaching is correct. To get a given letter (out of 25 possible letters) back with 50% probability, one already has to type 17.7 times on average, with two given letters it is 433 times, with 1000 letters already about 6037×10^{1397} strokes (Hehl 2016). However, the text of Hamlet comprises about 130,000 letters without punctuation.

These numbers are mathematics and have nothing to do with physics. Numbers in modern physics are also very large: the number of atoms in the universe is estimated to be between 10^{80} and 10^{82}, which is a very large number of physical significance. A very large number from IT for comparison is, for example, the number of instructions that a current supercomputer (10 petaflops) executes in a year: 3×10^{23} operations.

Thus there are several levels of "large" numbers with a typical environment and their own laws:

- In daily life, one million (10^6) is already *"big"* for us,
- for physics, numbers like 10^{23} to 10^{82} are large or *very large,*
- in contrast, numbers like 10^{1000} or "10 to the power of 10 to the power of 10" are completely different mathematical worlds with their own laws, but probably without any immediate physical meaning and without any associated countable objects.

The first level is the world of human numbers of objects, the second level are numbers, which at least in the lower range are still attainable for supercomputers, the upper range comes about by the combinatorics of large numbers and is no longer usual physics. Already the second level is effective infinity for us humans. The "real" mathematical infinity would then be level four!

The problem is the size of the search space, the set of all possibilities, if one allows all places, also the corners, so to speak. The technical term for this is "ergodic" from the Greek έργον 'work' and όδος 'way'. "Ergodic" means that in a system all corners are accessible and also visited by the many participants. However, an evolutionary mechanism does not need to go to all corners, but only to the right places in the landscape.

Already with the Hamlet text, one actually only has to consider a fraction of all points of the search space: It is, after all, English text, i.e. the letters are not all equally likely to occur, English words must be created, and English grammar must be observed - albeit according to Shakespeare's time. In addition, for all the poetic license, there are also the constraints of what makes sense in the first place. If, instead of pure chance, one would rather write a program that, in the style of Shakespeare, produces texts driven by chance on a continuous basis, then the scope of the problem would be even smaller.

The evolution works anyway mainly on the program level in the compact form of the blueprints. Therefore, in general, the situation is better if one does not want to create an optimal product directly, but to generate the associated optimal production method. To write a program that writes sonnets in the style of Shakespeare or composes cantatas in the style of Johann Sebastian Bach (Fig. 7.7) - this is not impossible today. The following example with the random typing of π shows it realistically: The program can be very compact.

Here is a realistic implementation for calculation:

"**Michal Majer improved his assembly language program**[2] **and achieved a tiny executable assembly language program of 121 bytes that computes the first 9280 digits of pi.**"
Downloadable from pi.com.

Thus, a program can be short and produce a long train of meaningful data. If one searches
the search space not for the output, the phenotypes, but for the generating programs, the
genotypes, the search becomes more compact; however, a small change to the program can
produce an arbitrarily large change in the behavior of the program. Often this means the
crash of the program run corresponding to the death of the individual in biology.

There are two measures to get close to the optimum despite the huge number of possible
combinations:

- A strategy on how best to proceed from a point reached and
- massive parallelism, i.e. simultaneous experiments in many, in very many places. We
  humans have a problem to imagine the huge dimension of parallelism of global
  evolution.

6.2.3 Directed Randomness and "Propensity"

Chance and the guidance of chance play a decisive role:

How to choose the initial values? And how does one go further? After all, our evolution
is moving searchingly in a high-dimensional space. But the good news is:

- We do not need to know any internals of the functioning of the solution,
- we do not have to be familiar with the search area.

However, we must also reckon with the fact that

- we do not find the very best solution.

One tries a random walk in a landscape to find the mountain top. Figure 6.6 according to
Hehl (2016) shows in the diagram (a) a random walk from point A to B. The two sketches
(b) and (c) indicate two path strategies to get from A to B: a method that searches
undirected to all sides and a method with a strategy that searches specifically and continues.

The (Fig. 6.6c) assumes that we can perform candidate search intelligently, following
some favorable "propensity". The word "propensity" here comes in this sense from
philosophy from the aforementioned Karl Popper and, even more anciently, from the
American philosopher and mathematician Charles Sanders Peirce (1839–1914). Fellow
philosopher Bertrand Russell called Peirce the greatest American philosopher in 1959.

---

[2] Assembler is a simple programming language designed for a specific type of computer.

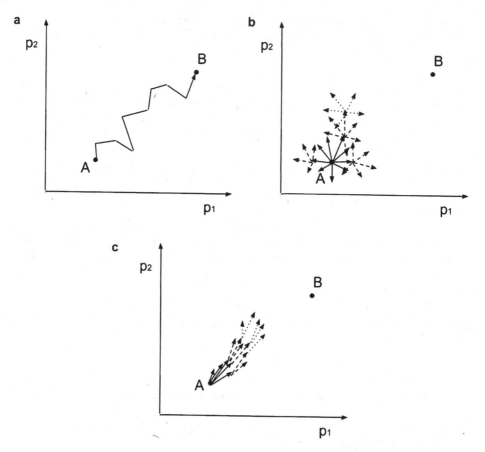

Fig. 6.6 A path in the state space: the actual path to the optimum B, the undirected and the directed search strategies. Image: own, Hehl (2016)

They called randomness with propensity as it occurs in the throwing of marked dice "propensity". Loaded dice already existed in antiquity and thus a first understanding of stochastics! For Popper (who already knows quantum theory) it is an inherent direction also of quantum mechanics, for Peirce it is a fundamental tendency of nature to break out of the rules and laws.

The Merriam-Webster dictionary defines:

▶ **Definition Propensity - an often intense natural inclination or preference, from the Latin propensus - 'inclined, hanging down'.**

What is meant is not chance, but the tendency to chance, and in particular to a special chance.

With the concept of "directed" chance, Popper imputes to the more or less random processes a kind of vague force that includes the real forces and chance in its effect. Physically, nothing changes, but philosophically, chance is valorized and becomes a component of the world.

We illustrate directed randomness by a directed random construction: the growth of a snowflake. If you like, it is an example of "propensity", in the sense of both Popper and Peirce.

It has been a problem in science for four centuries and has long been a mystery:

"Why do snow crystals, even before they become big snowflakes, always show six corners and six arms, like a feather?"
   Johannes Kepler, in "Vom sechseckigen Schnee", Strena seu de nive sexangula, 1611.

Figure 6.7 shows a perspective view of the lattice of atoms or of water molecules. Each oxygen atom is connected to two hydrogen atoms directly (via a chemical bond) and two indirectly (via a hydrogen bridge). The picture shows some planes of the area where the snowflakes grow and the six-ray symmetry of the structure. As the crystal grows, a water molecule from the cold, water vapor-saturated atmosphere outside reattaches itself in just the right way: hexagonal rings in the plane, tetrahedra spatially around each oxygen atom.

The growth of the flake is random, but with hexagonal symmetry. The process is causal, random and with hexagonal symmetry - a kind of attenuated causality with directed chance, a "propensity". Without this "propensity" there would be no flakes, but small balls of ice. There are impressive computer simulations for the growth process.

Searching with chance and a search strategy is a pragmatic general procedure called a "metaheuristic". The underlying verb is.

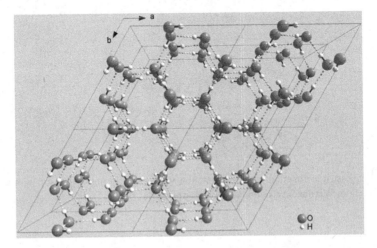

Fig. 6.7 Crystal structure of common ice viewed along the c-axis, perpendicular to the snowflake. The snowflake grows outward in planes. Image: MCryst struct ice, Wikimedia Commons, Solid State

εύρίσκω (heurískō) 'to happen by chance, find, discover, obtain'.

This includes the famous exclamation *"heúrēka"* - I found it.

A metaheuristic generally does not provide the absolute best and exact solution, only a "relatively best" solution. Proposals for procedures for directed random search have very pictorial names and ideas, such as.

"Mountaineering algorithm" like a climber looking for the summit in the fog, or *"simulated annealing"* like giving atoms lots of energy by heating them to get over a hill too and find another, better optimum.

A particularly difficult terrain for a semi-random odyssey and a search for the highest peak or for the deepest valley is a mathematical landscape with many peaks and valleys as in Fig. 6.8. The Russian mathematician Leonard Rastrigin invented this "testing ground" in 1974. It is a testing ground for evolutionary algorithms: Do you find local and supralocal extrema? Or get stuck in one region and not find the higher peak next to it? We will interpret the picture differently in the chapter "biological evolution".

However, the simple "brute force" methods as well as these more sophisticated "directed" random methods are only subsets and simple reflections of the great game, the evolution of species.

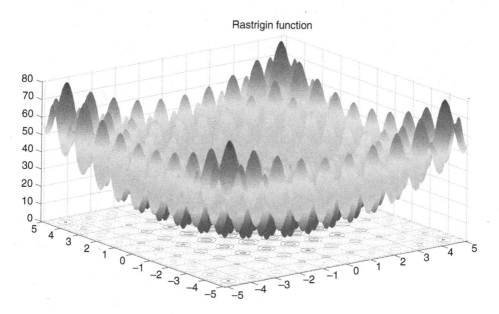

Fig. 6.8 The Rastrigin function as a test function with many local maxima or minima. Image: Rastrigin function, Wikimedia Commons, Diegotorquemada

6.3 Biological Evolution as Creative Chance

6.3.1 Charles Darwin

"Never has a doctrine established by a single man [...] proved so true as the theory of descent of Charles Darwin."
Konrad Zacharias Lorenz, Austrian zoologist and Nobel laureate, 1903–1989.

Probably never again has a single scientist had such a great influence on science and "neighbouring cultural areas"! At the same time, the scientific possibilities, which Darwin had, were very limited. Let us compare the following starting points at his time:

- the world of the "early creationists" as described above with human short times, human numbers like thousands of temporally constant, fixed species and a Creator King who vividly "does" everything,

and

- today's abstract conception of evolution with billions of years for development at stake, with many millions of species constantly changing, and with inherited DNA nuclei for everything at unimaginably tiny molecular scales.

If we look at the world of 1859 in biology and religion, we are amazed at the courage and scientific greatness of Darwin. Darwin wrote to the botanist Joseph Hooker in 1844:

"I'm almost convinced (quite contrary to initial opinion) that species - this is like confessing to murder - are not immutable."

The particular and striking evidence of this is the group of birds he observed (and shot) in the Galapagos Islands, which now bear his name: the Darwin's finches (Fig. 6.9).

He clearly observed a microevolution on his voyage with the Beagle in 1832, or rather its result: The Galapagos Islands are 1000 km away from the mainland Ecuador; probably a pair arrived on one of the islands as a founding pair by storm or on driftwood. From here the finches "radiated" to other islands with separate adaptations (so-called adaptive radiation). Today 18 closely related bird species are known. Darwin had described these birds in his diary and travelogue, but did not use them as an argument in the first edition of his work. Only the cooperation with the ornithologist John Gould he realized the importance.

But the birds and the trip to the Galapagos were and are a stroke of luck to see clearly and without high tech the change of species until the separation into separate species, after which no mating with the old species is successful anymore (reproductive isolation). Today's technology has identified the gene ALX1, which is responsible for the different beaks:

1. Geospiza magnirostris.
3. Geospiza parvula.

2. Geospiza fortis.
4. Certhidea olivaɔea.

Fig. 6.9 Four Darwin finches or Galapagos finches. From the Voyage of the Beagle, Darwin, 1845. Image: Darwin finches by Gould, Wikimedia Commons

"I wouldn't be surprised if it turns out that mutations that minimally alter the function of the ALX1 gene also contribute to the amazing diversity in human faces."
Leif Andersson, Swedish geneticist, born 1954.

6.3.2   The Concept of Evolution

In general, evolution (from Latin *evolvere* "to roll out", "to unfold", "to develop") is the gradual further development of a system, or better a system of many systems. Today, evolution is understood to mean the development of the biosphere as a whole, of course with man as the provisional end point. Evolution in the narrower sense is the reciprocal development of a system of participants with chance and a criterion for success. In addition to biological evolution, this term includes, for example, the development of technologies or literature. In biology as in economics, the criterion of success is growth, or at least survival. From Darwin comes the theorem:

"It is not the strongest species that survives, nor the most intelligent, but the one that responds best to change."

Biological evolution is probably the greatest spectacle of chance that we know, with the most diverse forms of life, including us humans, and thus of fundamental importance, also for us.

For Darwin, the word "evolution" was at first a suspect word, evidenced especially by a less scientific predecessor, the successful Scottish author Robert Chambers (1802–1871). Chambers had written the best-selling *"Spuren der Naturgeschichte der Schöpfung"* (*Vestiges of the Natural History of Creation*) in 1844 in journalistic style. This was a few years before Darwin's magnum opus, *"Über den Ursprung der Arten durch natürliche Selektion"* (*On the Origin of Species by Natural Selection*) in 1859. Chambers' book used and watered down the concept of evolution to any evolution - of the cosmos, stars, earth, animals and plants to (Caucasian) man. It was a collection of speculations, e.g., claiming the "spontaneous generation of insects by electricity." It provided no method for evolution itself. The book was eagerly read, e.g. by Queen Victoria and President Abraham Lincoln, but it provoked much criticism among scientists (and the Church). This was the reason that Darwin hesitated so long to publish it.

Darwin does not use the word "evolution" until the sixth edition of his *Origin of the Species*. But he, unlike Chambers, has a scientific method for evolution: variations in organisms produced by chance and then natural selection in a world with chance. Darwin writes in his *Origin of Species*:

"It follows that any being, if it vary however slightly in any manner profitable to itself, under the complex and sometimes varying conditions of life, will have a better chance of surviving, and thus be naturally selected."

Darwin calls this interacting life of organisms in the diversity of chance "sporting".

We distinguish today:

* chemical evolution, the very first development of the complex and copyable molecules from simple substances needed for protein-based organisms to evolve,
* biological evolution, sometimes divided into macroevolution (the big steps) and micro-evolution (small changes, e.g. what is directly visible in our lifetimes).

Microevolution can be seen or felt directly, such as the formation of the beak shapes of Galapagos finches or the frequently observed acquired resistance of bacteria to antibiotics. It is obvious.

Today we know that there is no hard distinction between micro and macro, such as species boundaries as a dividing line. The boundaries of a species have become permeable, in the laboratory recently, but always in nature. Macroevolution is the long-term integration of microevolution. Both humans and nature can insert DNA from one species into DNA from another. Bacteria in particular can also exchange genetic material between different species independently of reproduction processes. Typically, a few kilobytes to a few hundred kilobytes are inserted into the foreign cell.

In other words, evolution not only continues vertically with randomness from generation to generation of organisms, but also develops horizontally with randomness across species. Evolution is a two-dimensional random network of software blocks that usually become more complex and functionally powerful as they progress. But not always!

The laws of the development of large software systems apply, as in the software development of Microsoft, IBM or Credit Suisse (Hehl 2016). Vertically, these systems are further developed from program version to program version; horizontally, proven code is exchanged with other developers and reused.

6.3.3 Mechanisms of the Action of Evolution

Mutations

The task of biological evolution is more profound than, for example, finding an optimal wing shape alone. It first had to find the right "hardware" for life and finally the program that builds the optimal wing shape. The change of the blueprints, mostly deterioration and the occasional improvement, takes place through spontaneous mutations, for example through radioactivity and ever-present cosmic radiation or through chemical influences:

> **"I suspect any worries about 'genetic engineering' may be unnecessary. Genetic mutations have always happened naturally, anyway."**
> **James Lovelock, British physicist and thought leader of the ecology movement, born 1919.**

At least the second part of the quotation is unquestionable! Organisms are also like digital computers in this respect: After solar flares particle streams from the sun hit the earth and mutations and also errors in the electronic data memories of computers increase.

At the molecular level, the mutations can mean many things, such as an exchange in an amino acid base pair, insertion of an additional DNA strand, duplication of a piece of DNA, or a change in the number of repeats in the strand. It almost doesn't get more random than that. A cosmic particle from a supernova thousands of light years away from us creates a cascade of particles in the atmosphere and one of them is creative in us!

If the mutation takes place in a cell that later gives rise to germ cells, the change is transferred to the population and its gene set, the gene pool. A new trait is created in the population.

Recombinations

> **"God plans all perfect combinations."**
> **David Brainerd, missionary to the North American Indians, 1718–1747.**

Genes are mixed in a population through sexual reproduction. Whole chromosomes or parts of chromosomes can be recombined. The number of possible combinations is therefore astronomically high: humans have 23 pairs of chromosomes: there are 2^{23} possibilities for forming a single ("haploid") set of chromosomes. If the normal, double ("diploid") set of chromosomes is formed at fertilization, then you get $2^{23} \times 2^{23}$ possibilities: That's about 70 trillion variants for the possible randomness in an act of reproduction. This number also hides the directed coincidence in the choice of a partner with the preferences for the hair colour of the partner . . .

Selection

"Nature cares nothing for appearances, except in so far as they may be useful to any being. She can act on every internal organ, on every shade of constitutional difference, on the whole machinery of life. Man selects only for his own good; Nature only for that of the being which she tends."
Charles Darwin, Origin of Species, 1859.

Natural selection as a scientific concept comes from Charles Darwin: all hereditary traits that lead to a difference in successful reproduction rate affect the interaction with the other organisms of the same species, other species and the environment. The advantage or disadvantage inevitably, quasi deterministically, affects the coexistence in the population. As a result, the gene pool changes.

But chance also strikes populations on a large scale, rapidly with the impact of a large meteorite or even asteroid, with a volcanic eruption of global significance, or more slowly with a climate change that causes deserts to expand, for example.

What matters then is adaptability, the *"survival of the fittest"*, the survival of the individuals or species best adapted to the local environmental conditions. In the fifth edition of his "Origins of Species" Darwin adopted this so often misunderstood expression for natural selection, but not in the "sporting" sense, but in the sense of the best adaptation.

Population Movements and Chance

"All cheetahs alive today are as closely related as laboratory mice after a long period of inbreeding. A possible reason is thought to be that the cheetah population once went through a bottleneck, i.e. was close to extinction."
Universität Wien, Franz Embacher, Skriptum 'Bedrohte Arten, das Schicksal der Genen und der Zufall in der Evolution'.

In the mechanisms so far, chance has acted directly on individuals and changed them, but chance also indirectly changes the whole gene composition of a population, simply by its laws and distribution. Most gene changes in individuals are neutral to selection, neither beneficial nor detrimental, but there are still changes due to changes in the group itself.

Bottleneck Effect
If a disaster strikes a population, such as an epidemic or trigger-happy immigrants or a food shortage, and many individuals are removed as a result, this creates a random sample from

the population with a new, narrower statistical distribution. With very small samples, the statistical fluctuations with the remaining individuals now become drastically large and the sample has, in particular, a very special composition of gene variants that deviates from the initial population.

This is illustrated by the sketches in Fig. 6.12. Natural events reduce the population to the small circle. In the small sample, the green gene variant ("allele") is not represented. If the population has recovered by reproduction in the right circle, the variant distribution has become much more uniform and poorer. By chance, some gene variants may even have disappeared. One says: The gene pool has (or is) drifted.

Several species have undergone bottleneck developments or are now at such a sensitive and critical stage - after all, a realistic alternative is extinction, the disappearance of the species. An example is the cheetah (Fig. 6.10). Based on accurate genetic studies, it is estimated that the drama happened about 10,000 years ago. This would mean that all present-day cheetahs are descended from only a few individuals. All animals living today are thus so closely related that tissue can be transferred from one cheetah to any other without a rejection reaction, as if they were identical twins.

It seems that modern humans have also passed through a narrow genetic bottleneck. One hypothesis assumes a dramatic coincidence and a short bottleneck. The bad coincidence was the eruption of the Toba volcano in Indonesia about 70,000 years ago. According to another hypothesis, humanity's bottleneck lasted about 100,000 years. After that, there would have been few humans until the Paleolithic era. In any case, the history of mankind is a dramatic game of chance on both a small and a large scale.

Inbreeding is a deliberately created genetic bottleneck with an artificially limited supply of gene variants:

Fig. 6.10 Cheetah as an example of a species with low genetic variability. Image: Cheetah in the Ngorogoro Crater. (Source: A. jubatus, Wikimedia Commons, Rob Old on Flickr)

"If you look at most of the royal houses in Europe, the inbreeding has been quite extraordinary."
   Nikolaj Coster-Waldau, Danish actor, born 1970.

If in royal houses the focus is on political aspects, in animal breeding the focus is on the desired physical and character traits. However, as a side effect, the "good" chance that ensures genetic health is reduced and the existing unfavourable chance remains. In humans, for example, it is the potential of genetic disorders; today about 6000 different possibilities have been described and new defects are constantly being added.

Figure 6.11 shows some hereditary diseases that can be assigned to a gene on a specific chromosome. It is a big roulette game with many "chances to win" that we are all playing here, or rather have already played!

And it is not the only roulette in which we and our ancestors have played: The history and prehistory of our lives is coincidence upon coincidence upon coincidence.

Actually, we play Russian roulette all the time:

"A stupid game of chance played with a partially loaded revolver."

And many non-born people even played "Polish" roulette:

"It's like Russian roulette, but with an automatic pistol (so the first shot hits)".

Both definitions are from Urban Dictionary, pulled April 2020. We play these games as individuals to ourselves, as parents to our children, as the species *homo sapiens*, that is, as all of humanity.

Founder Effect
The transition from a small population to a large population (the right side of Fig. 6.12) establishes a settlement with a species. The transition from a large population to a small population can also be interpreted as a geographical "breakout", like the journey of the Darwin's finches from mainland Ecuador to the Galapagos Islands. The gene pool has become smaller, the remaining population more sensitive to disturbances. But the conditions for the formation of a separate species are better in isolation!

In physical terms, in the transition to a small population, entropy has been drastically reduced. When propagating with less possibility of variation, the entropy remains lower than in the initial state. So, in physical terms, a civil marriage increases the entropy of the royal house! High entropy is evolutionarily healthy.

This mathematical effect of chance is well known in statistics, albeit in the other direction, regression to the mean. To do this, one can read Fig. 6.12 from the small circle (i.e., a sample) to the left. If one first gets to know a small sample from a set, say a person from a group or one is given a single quiz question, and one finds this person likeable or one can answer this question brilliantly, the effect of the "regression to the mean" urges

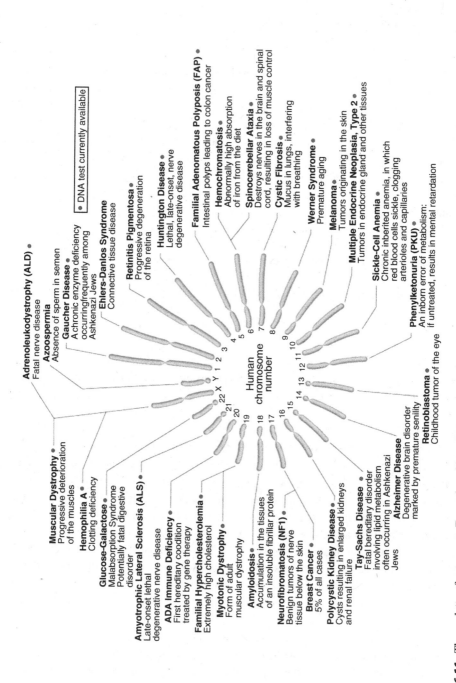

Fig. 6.11 The roulette of our genes. Examples of hereditary diseases per chromosome. Diseases that can be detected by DNA analysis have a red dot. (Source: Human Chromosome Diseases Set, Wikimedia Commons, Игор Петков)

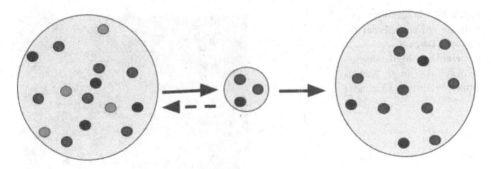

Fig. 6.12 Gene drift, schematic sketches of the bottleneck effect (left arrow to the right), founder effect (right arrow to the right) and "regression to the middle" (dashed arrow to the left)

caution. "Normally" the people in this group are not so friendly and the questions are not so easy to answer.

To conclude the chapter, a physical observation about a curious and perhaps interesting parallel to the bottleneck effect from a very different area of chance, the throwing of a stone into water described above. In the physical implosion, details of the past are also destroyed. In physics it is the energetic clash of energy in the implosion region, here the violent stochastic deviations from equilibrium in the stochastic implosion in the reduced set of individuals.

6.4 Anthropic and Copernican Principle

"The more I examine the universe and the details of its architecture, the more evidence I find that the universe in some sense must have somehow known we were coming."
Freeman John Dyson, English-American theoretical physicist, 1923–2020.

The physicist Dyson is extremely versatile; he successfully deals with quantum physics, mathematics and rocket propulsion, but also with astrophysics, science fiction, global warming and evolution. This quote is a witty paraphrase of the "anthropic principle". The term was invented by Australian astrophysicist Brandon Carter in 1973 as a counterpart to the "Copernican" principle (see below). The word from the Greek *anthropos* 'man' refers to us humans, the anthropic principle to our position in the universe.

The anthropic principle says tautologically that everything in the universe and our history must be so in order for us to observe it: If it were not so, then we would not be there either. We've already discussed it above, but it doesn't help in knowledge. First, the conjecture of Giordano Bruno (1548–1600) has more substance, the so-called "Copernican principle".

Figure 6.13 illustrates the "Copernican principle" as seen by the astronomer Johannes Kepler around 1600. Each star in the picture is a sun like our sun, and the letter "M" stands

Fig. 6.13 Giordano Bruno's universe in Kepler. From the *Epitome Astronomiae Copernicanae* by Johannes Kepler, 1618. Image: Kepler-Bruno, Wikimedia Commons

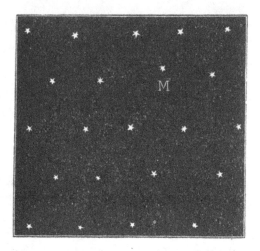

for *mundus* - for our world. The star M denotes the *stella mundi*, our sun. Our world is one of many and the universe is open, not a closed sphere as before. In another picture the stars are also irregularly distributed, i.e. by chance. Copernicus still had all fixed stars attached to the large outer crystal sphere, i.e. at the same distance from the sun.

Therefore, the principle rather stems from the conception of the natural philosopher and magician Giordano Bruno, who wrote - without any proof, as pure speculation:

- *"In the universe there is no center and no circumference, the center is everywhere"* and
- *"The countless worlds in the universe are no worse and no less inhabited than our Earth."*

Perhaps the truth lies between the two principles "Copernicus or Giordano Bruno" and "anthropic". We have a long chain of conditions for our life from astrophysics to the invention of intelligence. Some links in the chain look like they necessarily lead on, such as supernovae generating heavy elements for building versatile molecules. Some properties look like we just happen to be the lucky ones among many candidates in the universe, e.g. with the special habitable zone for our kind of life. Ultimately, there are conditions, or at least assists, that simply make our existence look like a unique coincidence. One example is the Earth's moon.

Our Moon was formed about 4.4 billion years ago when the young Earth was struck by a planet the size of Mars with the hypothetical name Theia. This led to two extraordinary and favourable boundary conditions for the development of life (Kleine 2019):

- The big moon has saved the earth from too much "lurching". The Earth-Moon system is much more stable than the Earth alone.
- The infalling planet came from the outer reaches of the solar system and therefore brought a lot of ice (water) with it. We owe our oceans to this event.

We cannot decide, whether everything is "nothing special" in the sense of "it is all just common, small coincidence". But perhaps there is a special one, a "big, unique" coincidence, which makes everything unique in the whole universe. Are we the only players who have come this far on the field? Are we that special?

To this again, a quote from physicist Freeman Dyson with a smirk from an interview a week before his death (Mack 2020):

"The beauty of science is that all the important things are unpredictable. The optimistic view in me is that nature is designed to make the universe as interesting as possible."

Another kind of optimism is that our history is such that we have existed so far, but also continue to exist. Not only evolution, but also the Stone Age, the Bronze Age, the Iron Age, coal-based industry, nuclear energy and climate change had to be and have to be, no way around it. After all, we can go on without nuclear war and with a sustainable economy - says the optimist.

6.5 Evolution: All Coincidence or Also Necessity?

6.5.1 The Eye as an Example

"The whole process from the first rhodopsin, the eye pigment, to the high-resolution visual organ took about 170 million years and was nearly complete at the onset of the Cambrian, about 530 million years ago. The evolution from shadow detectors to multi-directional photoreceptors has led to secondary developments in eye evolution in bivalves, fan worms and beetle snails."
Daniel E. Nisson, Swedish biologist, in "Eye Evolution and its functional evolution," Vis Neurosci, 2013.

The eye is a popular historical object of evolution by its wonderful technology and its visible perfection; it was already for Darwin the difficult object of "his" theory. Today it is clear that the eye is not a NO-GO problem. According to the German-American biologist Ernst Mayr, there are 40 independent evolutionary paths to the eye. The evolution of the eye throughout the animal kingdom up to man has become an established and elaborated part of evolutionary biology.

The difficulty of imagining the eye as the result of chance is also based on the idea that the complexity of the eye arises as a whole. But this is not the case. Even complex software systems consist of functional layers and many building blocks. Evolution had hundreds of millions of years to build the eye.

From an evolutionary point of view, the sense of sight is a relatively clear task, teleologically speaking: The physical task is clear, namely to transform light signals into information about, say, a prey animal or predator. This obviously provides an essential benefit to the wearer. Consequently, the sense of sight is a possible *convergence property* for evolution.

▶ **Definition** We define a property as *convergent* with respect to evolution if, given different initial conditions but the same external conditions, evolution would again find a similar solution.

Since evolution means multiple layers of matching chance until, for example, we could arise, this is not self-evident. If one assumes that a function would not evolve a second time because a suitable chain would not come together again, then we define this property as *contingent* according to the philosophical term for chance, namely contingency from the Latin *contingentia* 'possibility, chance'.

Figure 6.14 shows a modeling of the lens eye based on a classic paper by Nilsson and Pelger (1994). Starting from a light-sensitive speck, the optics of the human eye emerge in this simulation after about 400,000 generations. The numbers between stages give a measure of the difference in external features from the beginning of one developmental stage to the next. According to this work, the internal assumptions are chosen "pessimistically" (i.e. rather too difficult for further development) and in particular no hard problem, a "show stopper" for the development to the optically good lens eye has been found. The 400,000 or so generations corresponds to a geologically short time, orders of magnitude shorter than the time life has existed on Earth. However, with the eye only the sensory "hardware" of the sense of sight has developed. It is only the associated and appropriate neural processing and application that make the value of the sense of sight to the living being. And there remains the problem of the beginning, the small chemical evolution of the light-sensitive basic material.

6.5.2   The Big Question: Small Coincidence or Very Big Coincidence?

The question of contingency of evolution, chance yes or no, is a philosophical continuation of the astrophysical question whether there is other life in space on any exoplanets. That evolution has prepared everything so well on Earth, at least in the biological basics, is a fact. Evolution obviously works (de facto) by many small coincidences. The question is about the unique, very big coincidence: did it exist? It is first of all again an "anthropic principle", because we are there to observe it. But are we contingent, i.e. random-unique?

Would evolution repeat itself in much the same way if the same primordial earth were available again?

Or were there such improbable coincidences in "our" evolution that the new evolution would get stuck? Maybe with anaerobic bacteria, with algae or with dinosaurs?

Fig. 6.14 A classical simulation of the development steps of the lens eye. The numbers indicate the required simulation steps of the optical evolution. Image: Model eye Nilsson and Pelger, Wikimedia Commons, Gagea

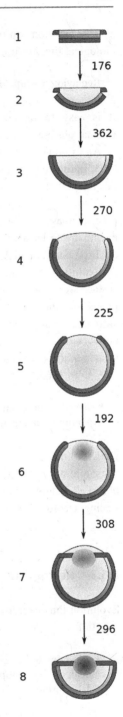

This question can be answered in the negative fairly confidently for many functions of biological life, but not (yet) for all. One article (Blount et al. 2018) briefly states the task:

Replaying the tape of life - Are we rewinding the tape of life?

It is easy to understand that evolutionary developments occur when the physical foundations are available or easily attainable and clearly linked to an evolutionary benefit.

An example is the shape of the dolphin body with the very small drag coefficient C_w of 0.03, optimized for fast swimming in water. Any body for which speed in water is an advantage must necessarily assume a similar shape.

Also the eye with lens rather belongs to this category. A device that is supposed to give a precise image of the environment with light will come to curved surfaces and thus eventually to lenses. The life advantage is obvious and tends to grow with better image quality - and new niches open up in the feed market!

Evolution has a number of general optimization tasks to solve, such as the appropriate surface area of leaves in plants, small in desert plants, large in shade plants. In animals and plants it is the distribution and collection of fluids. When simple diffusion is no longer sufficient because the dimensions become too large, the common solution is a hierarchical network of veins and capillaries. For leaves, it is a matter of optimizing the mechanical system of the stem and branches on the one hand and the leaves on the other. Darwin wrote already in 1865:

"The advantage gained by climbing is to reach the light and free air with as little expenditure of organic matter as possible."

Details of the leaves are random, but the overall structures are similar to minimize material consumption, maximize stability, and achieve optimal certification by the sun. As Galileo observed, nature can enlarge or reduce the same structure, but not simply linearly, but with scaling according to its own laws (Fig. 3.12).

6.5.3   Retrograde Directed Randomness

Evolution thus exercises "directed chance", at least in the sense that chance is tested by selection after the event and cancelled if necessary:

"Continuing to live gives meaning to a certain coincidence for the organism - or not. The coincidence just happens. It happens."
   Klaus Mainzer, German philosopher, 2007.

Or more vulgarly, if the coincidence turns out badly, "Shit happens."

The directed coincidence in the case of the snowflake was forward-directed and immediately constructive, but the mechanisms of evolution also act like a "Popper's propensity", albeit retrograde, in retrospect. For science, biological evolution has a great advantage over political and human-historical evolution: one can replay the "tape of life" in small pieces in the laboratory, e.g. in bacterial cultures.

This can be done experimentally with the famous *E. coli* (named after the Ansbach physician Theodor Escherich, hence the E.). With bacteria, the generation sequence is fast and one can even freeze cultures and reuse them later for later comparisons.

The researcher can play evolution on a small scale and let it develop in many different ways, for example with

- several identical populations in identical environments,
- identical populations in different environments and
- different populations in the same environment.

Probably the most famous experiment of this kind is the long-term evolution experiment LTEE of the Universität von Michigan. 66,500 generations of *E. coli* in twelve lineages were followed over 20 years. Eleven lineages evolved similarly, while one lineage evolved quite differently - it had evolved a new ability to process citric acid as well. The analysis revealed that a single mutation had accidentally prepared the way for this. Most of the evolutions were convergent, but not all. But even the convergent populations showed subtle genetic differences in the mutations present and signs of splitting into a second species.

The small, positive steps of microevolution are reproducible with high probability. Figure 6.15 symbolizes this by the many maxima, each of which is supposed to represent a relatively stable species: Convergent evolution leads to one of the stability peaks or around the peak area.

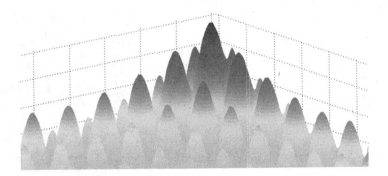

Fig. 6.15 The stability peaks in the landscape of species (symbolic). Image: Section of the Rastrigin function, Wikimedia Commons, Diegotorquemada

6.5.4 Difficult Beginnings and Megatrajectories

The most difficult situation and most sensitive to the right coincidence are the beginnings
of chains that eventually lead to a meaningful function. Most difficult of all (probably) was
the beginning of everything, of evolution itself. An IBM jargon word for this is *bogey*, an
expression that originally comes from the game of golf:

> **Bogey n**. *A goal, especially a difficult or unpleasant one. Used in planning cycles when people
> or budgets are cut. The new goal is then a "bogey" and means the unpleasant target that must
> be achieved within the specified limits. Echoes the way pilots spoke in Second World War.*

For the beginning of a functional development, it is most difficult to obtain an evolutionary
"reward" for the innovative achievement, unless the new function can already be started
quite simply like the eye as an inconspicuous but already useful light-sensitive spot. The
associated material innovation, a light-sensitive substance, is the rather complex compound
rhodopsin made of a protein and a color-bearing molecule, a relative of vitamin A
(Fig. 6.16). The figure makes it clear that rhodopsin is a substance that has also already
required evolution.

In the case of some evolutionary milestones, there is a suspicion of "contingency", i.e. a
one-off coincidence, and subsequent divergence, i.e. the divergence of characteristics in
different ways. These are developments with difficult feedback of the first successes, but
then great subsequent spread of the invention as a "megatrajectory". This is, for example,
the emergence:

Fig. 6.16 Illustration of the
complexity of the visual
pigment. Structural model of
rhodopsin. Image: Bovine
Rhopsin, Wikimedia Commons,
Jähnichen

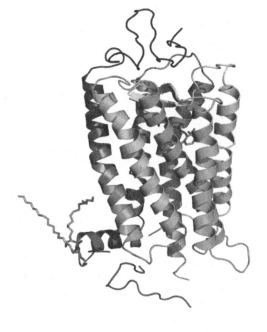

- of life itself from chemistry, called abiogenesis,
- of protective cell walls,
- of multicellular organisms of higher efficiency,
- of photosynthesis to obtain energy from the sun,
- of sexual reproduction to increase variability,
- of intelligence for better adaptability.

Figure 6.17 illustrates the development of evolution schematically from the standpoint of algorithmic complexity. The complexity of the organisms increases, since, in general, higher complexity also means better functionality (exceptions see below). The measure for complexity is chosen in such a way that there is approximately a linear growth. In the language of software technology this would correspond to the measurement of complexity in function points, the occurrence of new functionality. Then we find periods with very little growth (the curve is flat), with rapid growth (steep or even jumpy curve), and "normal" periods. So it could be that a new macro-step from the above list needs many attempts and a lot of development time until "it succeeds"; after that there is rapid advancement. Periods of high pressure on living things, such as the impact of the Yucatan meteorite, could also subsequently trigger a surge of innovation.

The series of trajectories of life could be continued beyond biology into the present time in the cultural sphere, for example with the invention of writing, printing, the computer and the associated creation of software by man himself, including the generation of artificial intelligence.

American astrobiologist Christopher Chyba, born in 1960, said,

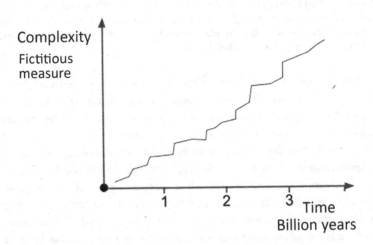

Fig. 6.17 Principle sketch of the growth of the algorithmic complexity of the system of biology over time. Image: own/ Hehl (2019)

if someone gave him an exact copy of the Earth 4.2 billion years ago, he wouldn't know if life was certain to originate or only with a 1 in 100,000,000,000 chance (one in a hundred billion). Yes, we know it happened here - but could that have been extremely flukish? Washington Post, April 4, 2014.

This is an expert, if unsatisfactory, answer to our question above about whether evolution would repeat itself.

We have already mentioned the "megatrajectories" in the deeper astrophysical past of life. There it is clear that the sun is not a special star, but the earth is special. The "habitable zone" in our solar system is at least "a little" extraordinary, the big moon of the earth is "quite extraordinary".

Each stage of evolution thus carries with it a random probability of successful passage through the stage, depending on the state of knowledge and the associated context. Each stage could make us something quite unique in the universe, not just special, as we undoubtedly are.

Evolutionary experiments show that our fate could hang on a single (random) change in a molecule.

6.6   Abiogenesis and Chemical Evolution

"Thus, he [the Spirit] turns the sweat of women and dogs into fleas and lice, and the dew into locusts and caterpillars, the glue into eels, the earth into plants, the carrion into worms, the excrement into beetles, besides infinitely many new and unusual things." Johannes Kepler, letter to David Fabricius, 1605.

Until two centuries ago, it was simply clear that at least simple living beings arise spontaneously and inevitably. The above passage from a letter by Johannes Kepler is instructive here! The question of the origin of life is not only scientifically interesting, but also philosophically and from the point of view of religion.

Until the seventeenth century, it was quite normal to assume the emergence of at least simple life spontaneously from decay. "Simple" life was considered by believers to be living beings without souls, such as the not-so-simple organisms in the quote above. Observation of single-celled organisms and microbes after the invention of the microscope seemed to confirm this. These organisms seemed to be primitive and so sexual reproduction (as a sign of higher development) was inconceivable in them. At the latest since the experiments of Louis Pasteur around 1860 it became obvious: even the simple life of microbes does not arise spontaneously, i.e. not just necessarily. The English biologist Thomas Huxley made the problem clear when he introduced the concept of abiogenesis, the emergence of life from dead matter, from inorganic or simple organic chemicals.

This observation strengthened the ideas of the Abrahamic religions and the idea of a Creator who had created all living things simply and once and for all. With scientific evolution the development of species becomes obvious, but there remains - religiously as well as scientifically - the question of the beginning of everything. One idea already comes from Charles Darwin. It is the idea of the famous *"some warm little pond"*. He writes in a private letter very carefully:

> **"But if (& oh what a big if) we could conceive in some warm little pond with all sorts of ammonia & phosphoric salts, light, heat, electricity &c present, that a protein compound was chemically formed, ready to undergo still more complex changes, at the present day such matter would be instantly devoured, or absorbed, which would not have been the case before living creatures were formed."**
> **Charles Darwin in a letter to Joseph Hooker in 1871.**

It looks as if Darwin was not entirely wrong with this assumption, which he only cautiously expresses. However, the progress of the sciences involved, and especially the newly emerged astrobiology, cell biology, virology and molecular biochemistry, has made the complexity of the task of "creating life from simple chemistry" really clear!

The formation of the basic organic substances of life from inorganic material has been demonstrated in various environments, such as in a primordial atmosphere with electrical discharges (the Urey-Miller experiment), in the vicinity of hot springs on the ocean floor, or even in different areas of space, from dark clouds to analyzed meteorites.

One promising direction of first abiogenesis, or chemical evolution, is the emergence of life in the hot, mineral-rich environments of seafloor "hydrothermal vents" nearvolcanic activity (Fig. 6.18). These are environments reminiscent of the first phase of Earth's history still without an oxygen atmosphere. Various chemical reactions are conceivable as energy sources, e.g. the transformation of iron sulphide with the formation of hydrogen sulphide or the water absorption of rock such as olivine with the formation of hydrogen and methane ("serpentinisation").

The idea of "panspermia" by the Swedish chemist Svante Arrhenius (1859–1927), which is already more than a hundred years old, attempts to explain life on earth through germs that have come to us from outer space. However, this does not solve the problem of chemical evolution, but only shifts it into space and thus gives it a science fiction character.

In our post-Popper model of the world, the emergence of life stands at the interface of world 1 (physics) and world 2' (computer science). In this sense, abiogenesis has two basic tasks to solve.

From the point of view of physics, the task is to make free energy available for life processes. The basic physical task is to produce hydrocarbons (such as methane) and other organic compounds from carbon dioxide under the conditions of the terrestrial primordial atmosphere and to create the energetic conditions for life. Adherents of the *"metabolism first"* direction consider this to be the first driving force.

Fig. 6.18 A potential environment for the origin of life. White fluffy mats at a submarine hot vent. Image: Champagne vent white smoker, Wikimedia Commons, NOAA

The basic informational task is to create structures that can copy themselves. The first candidate for such a substance is the versatile ribonucleic acid RNA, which can both transmit information and trigger various catalytic reactions. The long-term storage of collected information is then taken over by the much more stable DNA. This thesis is the *replicator first* doctrine.

From our anthropocentric point of view, the great ultimate goal is finally to build the tower of world 2, of computer science, with us humans and our intellectual products in the top position.

It doesn't help. On the question of chemical evolution or abiogenesis, most biologists become philosophers with different views. Tendentially, the hypotheses are plausible and argue for the "small" coincidence, which "usually" should lead to life. But nobody can exclude today the big unique coincidence for the big astrophysical-chemical-biological chain to us humans. Thus, the Copernican principle is weakened:

"[The world, life] is nothing special" - most of it, but maybe some of it is!

Astrobiologist Stephen Blair Hedges gives a little help in answering this when he says (2015):

"If life arose relatively quickly on Earth, then it could be common in the universe."

Perhaps Mars, with possible preforms of life, will provide clues here. That is at least one of the motivations for the current Mars missions.

From today's perspective, the oldest microfossils from oceanic vents are between 3.77 and 4.28 billion years old; the oceans themselves formed about 4.4 billion years ago. From the 13.8 billion years of the age of the universe, this looks like a short formation time! But it is likely that there was an epoch of violent first life developments, traces of which are hard to find.

6.7   Evolution as a Randomly Driven software System

"Competition is for losers. If you want to create and maintain lasting value, look to build a monopoly."
Pierre Thiel, co-founder of PayPal and Palantir.

The comparison of biological evolution with the evolution of digital software systems has inherent limitations. There are differences between biology and digital evolution:

- In biology, the individual is an active participant and not just a passive copy like the usual software product.
- In digital IT, the software product is developed *a priori* according to a specification, i.e. teleologically or "top-down", from above. Evolution gives the new functions their meaning *a posteriori* or it withdraws them by the carriers disappearing "from the market".
- The space of possible commands for constructing programs is well-defined in IT, rather chaotic in evolution.

However, digital customers also give feedback on their satisfaction and on errors and are thus not just passive. The subsequent bestowal of a kind of specification in evolution or sense-making - when one recognizes the effect of an innovation - is not so different from a preceding specification, which is then not strictly adhered to at all!

"Since when is computer software what people want? It's simply a matter of evolution."
Bill Gates, American entrepreneur and programmer, born 1955.

Indeed, much of the development of technology is driven by the technology and its possibilities, not by the desires of the "users". This is also expressed by the famous saying of Henry Ford (1863–1947), the automobile manufacturer: *"If I had asked people what they wanted, they would have said 'faster horses.'"* This is especially true of the great new breakthroughs, the megatrajectories. Of the many similar sayings, one more prediction, here from Ken Olsen (1926–2011), then president and founder of the once great computer company DEC (Digital Equipment Corporation) in 1977: *"There's no reason anyone would want a computer in their home."* It was the time of big computer boxes and data centers that ran corporate administration.

The situation is different for small requests for improvements and changes to a product. There is typically a long list of small and large requests and problems at any point in the life of a product.

6.7.1   Evolution as a Large System

"The current estimate of the number of species is between 5.3 million and 1 trillion."
Tanya Latty and Timothy Lee, April 2019, theconversation.com.

Living nature is in any case a very large, interconnected software system. The larger a technical software system is, the more similar it becomes in behavior to a biological system: it contains more and more randomness. More and more happens in the system, which even the developers do not understand or understand only with difficulty - even if all individual program steps are elementary and typically change and move only a handful of bits. A large software system is full of randomness: due to errors in building and running, due to users and their behavior and their desires. It becomes a distributed system that evolves as it runs. There are private software libraries in companies and there is software that is shared between companies. Some bugs that occur are solved "quick and dirty" by patches in the running code ("patched"), better fixes and newly developed features are collected and delivered as a package, as a new version.

All living organisms are - by definition, so to speak - running software packages with a lot of internal and external randomness.

Physically, running a program in the computer means passing on the spark of life to the next step in the process, probably neatly timed by the computer's clock. Even when doing nothing, there is one step after the other, some energy is consumed and the running system remains physically outside the thermodynamic equilibrium.

In the evolution of a software product, the product "lives" as long as customer-reported bugs are still being corrected. The name of the product - "Windows" or "WhatsApp" - is the overriding spark of life and a value. Under the name in the product, the technology can change, the name lives on as long as possible. This is similar with the higher-level evolutionary system, a company. An example here is the company Nokia: products of Nokia Oyj were in chronological order.

- *Paper products,*
- *Rubber boots and bicycle tires,*
- *Mobile phones, also as a world market leader,*
- *Communication networks and mapping services,*
- *Network equipment suppliers and technical special products.*

The employees are different, companies have been bought or sold. What has remained? You will probably say: the spirit of Nokia. A small part of the "Nokia" gene has remained over the years, such as *"Finnish company based in Espoo with the name Nokia"*.

The most important thing has obviously been

1. **Continue to live at all costs and**
2. **To be as fit as possible and/or as prepared for change as possible.**

Point 2 contains two formulations of Darwinism, in the popular form especially the slogan *"The survival of the fittest"* by the English philosopher Herbert Spencer (1820–1903). Nokia, by the way, in its greatest heyday, was as fit and as adaptive and innovative as could be imagined. They had established all the classic innovation mechanisms for companies (the author was somewhat integrated), but then still failed at a new megatrajectory. The mobile phone product mutated from a phone to a multi-purpose computer with millions of apps. The Nokia management noticed it too late and (somewhat exaggeratedly formulated) continued to build only good phones.

It is also no coincidence that there are systemic similarities between biological natural evolution and the evolution of large artificial software systems. These similarities are recognizable if one considers the observed laws from the software industry, the so-called Lehman's laws. The software technologist and IBM colleague Meir Lehman (1925–2010) made observations on the software evolution of large systems in large companies between 1974 and 1996. The software expert Madhavji summarizes the uncertainty to the principle of uncertainty in the world 2', the software world:

> **"No matter how many times a software system has performed satisfactorily, satisfaction on its next execution is uncertain."**
> **Nazim Madhavji, American software technologist, Madhavji** (2011).

Mind you, Nazim Madhavji says it for the technical digital world that seems so safe. But everyone has probably experienced that their PC stopped working after a harmless update. In the digital, the above statement is a bit extreme, because of course there are reliable software systems. A famous example is the computer system of the Space Shuttle, which was designed to be particularly robust - both mechanically and in terms of software (Fig. 6.19).

For safety reasons, there were five computers on board, four of which were redundant - the fifth was the back-up and responsible for take-off and landing. Everything possible had been done according to the state of the art in software technology to ensure safe operation. The system had to function reliably and correctly even after an error had occurred.

Lehman's software laws are not laws in the sense of physics, but rather something like "laws" in the social sciences, e.g.:

Fig. 6.19 NASA's emblem commemorating the Space Shuttle Program. Image: Space Shuttle Program Commemorative Patch, Wikimedia Commons, NASA/Dumesnil

- **Continuous change**: A software system must constantly change and adapt, otherwise it degenerates and becomes inefficient.
- **Increasing complexity**: The complexity of a developing software system evolves exponentially to its lifetime, at least as long as it is not maintained or deliberately reduced in complexity.
- **Self-regulation:** Software development as a whole is a multiple feedback process and must be treated as such.

The similarity and applicability of these sentences also to biological evolution and the biological system are obvious. These images are for many at first only metaphors, but they are more!

The following remarks from the description of the characteristics of the environment of software development of very large systems also fit biological evolution (after Northrop 2006):

Decentralized development and operation; work under inherently contradictory conditions; further development in operation; experiments with a wide variety of capabilities that are also abandoned again; heterogeneous and rapidly changing elements; development of new control mechanisms; failures in development are the rule rather than the exception (everything never works perfectly).

Northrop speaks of *"wicked problems"*: Every attempt to find a solution changes the system. In evolution, every changed or new species changes the world. This is especially true of us humans.

6.7.2   Evolution as Agile Software Development

"Agile development methods do not require any specifications. At the beginning, the customer only specifies a few basic functionalities and the project starts immediately." Homepage it-agile.de, pulled August 1, 2020.

The specification of a project describes the goal of the project and thus its purpose. So-called "agile" projects (lat. *Agilis* nimble, agile) have no end goal described in detail and no predetermined end product. Evolution does not have a requirements specification either! A software project requirements specification only makes sense if the task and environment of software development remain sufficiently stable. "Agile" development tries to keep all sub-processes as simple and agile as possible in order to be "agile". In view of the rapid development of information technologies and digitalization, it is not surprising that the vast majority of development projects today are carried out in an "agile" manner.

It is striking to see how nature realizes the principles of good agile programming and development. The rules for the agile programmer (Hehl 2016) can be found in the way evolution works, for example:

- Start without a fixed target,
- get to executable software as quickly as possible,
- quickly incorporate change requests and do not reject them,
- close cooperation "as a team".

The term "team" is to be understood comprehensively with the own species on the one hand and the other organisms, e.g. the prey animals, on the other hand. The essential thing is the functioning and thus the measure of progress. There are many failures, in nature even very, very many, but all in all something functional new emerges. The saying of the

American computer scientist Alistair Cockburn (born 1953) applies to the programmer as well as to evolution:

"We do things wrong before we do them right."

Overall, we can characterize evolution, that is, the global software development of nature, by observation:

Evolution is agile software evolving by chance itself over 4 billion years.

The analogy goes on into many details. For example, evolution likes to use program code that has already proven itself over and over again, even if it is not the best solution. This programmer's wisdom fits in with that:

"The best programmers never rewrite a program when they can use an old one for a new task."
Gerald Weinberg, American computer scientist, 1933–2018.

It is well known that proven code parts are used in industry for a long time and run in production, even if the designer no longer exists and no one understands the program. In the financial sector, it is therefore still COBOL programmers who are in demand, i.e. specialists for a programming language from the early days of computers 70 years ago!

However, if the quality of a program becomes too poor over time after a lot of patching, then the industry starts all over again, often under the blanket of the old brand name. But this also makes evolution with branching into a new lifeline. Algorithmic complexity reduces briefly at such points until it grows again through further development. Let's hope such a reboot doesn't happen to the human species!

Another wisdom from the IT industry and marketing also applies in evolution: *The winner takes all* - the winner gets (almost) everything. This statement applies to an entire species, to a male in a pack as well as to application software.

6.8   Religiousness, Evolution and Chance

"When Mother Nature speaks, even the gods hold silence."
Abhijit Naskar, Indian-American neuroscientist and author, born 1992.

Evolution itself is altogether proven, as it was correctly conceived by Charles Darwin without high tech. It is a central pillar of modern science. But it is an abstract system whose magnificence only becomes apparent when one realizes the immense dimensions of this development in time and scope. The grandeur becomes especially clear when compared to the anthropomorphic stories of creation, such as the Genesis of the Old Testament. It is science versus literature and symbolism. But the human concepts are simple, seductive, and firmly rooted in our psychology and intellectual history. The paradigm shift was (or is) not easy.

The best known theologian and evolution evangelist (with idiosyncratic theological and philosophical additions) is the French Jesuit and philosopher Teilhard de Chardin, 1881–1955, already quoted above. He drew a great world-historical optimism from the doctrine of evolution. Thus, he understands evolution as a further development of man with an unexpected outcome. His aim in life is the unification of evolution and Christian doctrine. But the Abrahamic religions cannot accept a further development without further ado, because we human beings are firmly convinced that we are the crowning glory of creation! That is why his life was also a single conflict with the Catholic Church and his Jesuit Order. De Chardin had already lost his chair in 1920, and he was not to live to see the publication of his major work, which was completed in 1940. It was only after his death that his books could be printed, and they quickly reached millions of copies. Today, Pope Francis has been asked to rescind the 1962 ecclesiastical reprimand of Teilhard de Chardin (Kathpedia, accessed June 25, 2020).

The classical occidental and oriental alternative to evolution is creationism with its literal interpretation of the Bible. Figure 6.20 demonstrates the absurdity of this doctrine with the "replica" of Noah's Ark. According to media reports, the exhibits also include chronologically inappropriate dinosaurs and fictitious biblical unicorns.

But modern science also has two philosophical-theological Achilles heels:

Fig. 6.20 The fictitious "Noah's Ark"at the Ark Encounter in USA. Symbol of naive early creationism: dinosaurs, the flood and humans together. Image: Ark Encounter, Wikimedia Commons, OlinEJ

- It is the yet unresolved critical ramifications in the timeline of evolution, especially the beginning (which is understandable, of course, since that is where the first beginnings of the new must be found),
- it's all chance, which means you can't get ahead causally. You have to be humble.

It is obvious that the branching points are particularly difficult to prove with fossils and also to understand.

It is still possible to think: At the branching points (a) God helped and e.g. introduced sexuality into evolution. But this thought is no different in depth than the idea: *"God has just saved me from the accident"* or *"God has helped me in the A-levels"* - both statements cannot be refuted, but they are also without any factual added value, although perhaps with psychological satisfaction. However, the believer will also have to say to himself *"God sent me this illness as a punishment or a test"*; the thought goes for better or worse.

Let us imagine the extreme that God directly controls the whole evolution, also the microevolution in every single step! This leads at first to the problematic thought that also the dark sides of evolution are explicitly caused by God - however, this is nothing theologically-philosophically new. It is the problem of theodicy, from ancient Greek θεός theós 'God' and δίκη díkē 'justice', i.e. "justice of God" or "justification of God". There is obviously injustice and evil in the world in contradiction to the image of the 'benevolent God'. For the individual, this is the biblical story of Job who experiences suffering as a test. For humanity as a whole, correspondingly, would be to see a pandemic as a test or punishment for the human species.

However, if all evolution is an action of God - trillions and trillions of times a second all over the world - then this doctrine turns into pantheism: God becomes identical with the world and God is chance, both small and great. Chance becomes God and God becomes chance. Then God would be another word for chance. Albert Schweitzer, German-French physician and theologian (1875–1965), expresses this kindly:

"Chance is the pseudonym the good Lord chooses when he wants to remain incognito."

In this sense, God is the "quasi-person" behind chance. Thus, God is the superset of the set of natural laws in the world plus the set of randomness generated by him in the world - but it is "technically" just a name without any added value.

The concept of a Creator who only takes over the management of critical phases in evolution has at least two problems: first, it is logically unsatisfactory (where is the boundary between "divine" and "normal" chance?), and second, it is depressing for believers as a "God of the gaps".

Logically it is not a consistent solution: The concept of creation generally tries to solve a problem by postponing the solution of the task "creation" to a mystical intelligence, which by definition cannot be explained. But you can get just as far with chance! Actually one would have to ask recursively always further:

Who designed the designer? Who designs the designer of the designer etc. in infinite regression.

This is our first hidden infinite regression, the second hidden regression later in free will with the homunculus effect.

It is not like in mathematics, where in a regression the residual error that is pushed away becomes smaller and smaller, for example in the sequence $1 + 1/2 + 1/4 + 1/8 + 1/16 + 1/32 + 1/64 +$ etc. etc.: Here the total "etc. etc." is only $1/64$ in total and can be made as small as you like. This is different in the logical chain above, which pushes the designer and creator further and further, step by step backwards: But the task always remains the same, it rather becomes bigger and bigger!

Why did the intelligent designer allow so many failures of evolution, the fossils of which we have as evidence, when he could have introduced the perfect with a "swipe" (in software with a "patch")? This thought corresponds to the problem of theodicy in morality in society mentioned above, namely, "Why is there evil?"

The Intelligent Design form of selective intervention at critical points in evolution is theologically a typical God-of-the-gaps idea, an idea that God is to be sought where there are gaps in science at a particular point in time. This is illustrated by the famous and much-used cartoon "You should be a little more detailed here" (Fig. 6.21). The image is used from Brexit to psychology and here to evolution. In evolution, it fits the final gaps - where "*an intelligent designer did that*" is not an actual, explicit explanation. It's another word for "*miracle*". A natural, conceivable coincidence as an override is the same in effect, but without mysticism.

After Darwin, the lens eye was such a gap and "bogey" for explanation for a hundred years. Today, the evolution of the eye in the animal kingdom, with many special features - such as color vision, night vision, farsightedness, underwater vision, or independent vision with two eyes - is an established science with sound knowledge. However, it is not clear whether there is a common predecessor for the eye or whether essential features such as the lens, retina and photoreceptors evolved via separate pathways and converged to form the eye.

For many theologians, religion's retreat to the unproven and unresearched is an extremely unsatisfactory behavior (Hehl 2019). Already in 1893 the Scottish evangelist Henry Drummond (1851–1897), inspired by the idea of evolution and Darwinism, wrote (Hehl 2019):

"There are reverent minds who ceaselessly scan the fields of Nature and the books of Science in search of gaps—gaps which they will fill up with God. As if God lived in gaps? What view of nature or of truth is theirs whose interest in science is not in what it can explain but what it cannot, whose quest is ignorance and not knowledge, whose daily dread is that the cloud will lift . . ."

"I THINK YOU SHOULD BE MORE EXPLICIT HERE IN STEP TWO."

Fig. 6.21 The famous Sid Harris cartoon: "You should be a little more elaborate here!" Here with us: on Intelligent Design and the gaps in knowledge. Image: Sidney Harris, ScienceCartoonsPlus.com

The German Lutheran theologian Dietrich Bonhoeffer will write the equivalent in 1944. He had read about contemporary quantum physics and had become aware of the problem of faith through advancing science:

> **"[If] the frontiers of knowledge are ever pushing out, God is also ever pushing away with them, and is accordingly on a continuing retreat."**

In any case, so-called Intelligent Design is far from science, and far from the Bible.

Evolution gives the sober technical basis. It develops like a stream, based on the laws of nature and with a lot of chance. It is possible to imagine more behind it, such as a driving force and a meaning - the sober technical development of evolution does not say more. Point.

The American theologian Gordon Kaufman (1925–2011) and "Professor of Divinity", i.e. professor of theology at Harvard, tried to draw the consequence from this inexorable evolution by chance; he wrote (Kaufman 2001):

"Creativity happens: this is an absolutely amazing mystery".

Since our world seems to work, he speaks of the "serendipity[3] of creativity" as a synonym for the concept of God. This God does not work through chance, but is chance. He turns what is actually a disturbing property of chance, namely to be unfathomable by definition, into a religious one, into a mystery.

God is thus the sum of the laws of nature and chance itself, not a being behind them, not the Creator, Lord or Father. In theological classification Kaufman is a pantheist (i.e. the abstract God is identical with the world), for comparison Albert Schweitzer is a panentheist (i.e. God is in the world and outside the world).

The positive attitude towards the world expressed in the concept of serendipity, happy coincidence, brings Kaufman close to the German philosopher Gottfried Leibniz (1646–1716), who wrote:

"We live in the best of all possible worlds. Everything that happens is good."

Leibniz received a lot of ridicule for this paradoxical claim, and Kaufman can't get his way theologically either; his thoughts are too abstract. What about human creativity, which also works with chance? Is our creativity also divine? At the latest with Beethoven, Shakespeare and van Gogh and their ilk, the question is no longer blasphemous.

The meaning of evolution goes far beyond biology. In it, everything comes together: physics, our physical existence, our spiritual being and the power of chance.

A nice concluding saying on "evolution" and "religion" comes from the Swiss aphorist Walter Fürst (1932–2019):

"I'm afraid God is a follower of Darwin."

6.8.1 Summary of the Chapter

Chance is everywhere, rarely as big as an asteroid impact, but small and microscopic and mostly even irrelevant. Who cares about the fine structure of the pebbles on a path or the leaves on a tree! Not so in evolution. There it is mainly the small coincidence in the genetic material and in the cells that has a big effect. Every rearrangement in the molecules of the

---

[3] For the concept of serendipity, see Sect. 7.1.2 "Creativity".

chain is an event. Evolution takes place above all in the construction plans of the world 2′ and thus multiplies the effects:

Evolution is the story of world 2′ with chemical evolution at the bottom and us (and computers!) at the top.

It is one great story of chance. Even the earth, even world 1, has been changed by evolutionary chance with the production of the oxygen atmosphere by the plants from world 2′. Today there is a wealth of scientific information on evolution that shows its intellectual magnificence compared to the childish and naive smallness of young-earth creationism!

But one must also not approach evolution naively. The typing monkey is not an option, this picture is pure mathematics and has nothing to do with reality. There is a certain propensity in chance for statistical processes, like how snow crystals grow with chance and hexagonal symmetry. We call it "propensity" according to Popper. This propensity reduces the size of the task.

Technically, evolution is a software-technological method for creating something new. In principle, we in IT have copied the principle from nature. This is actually something unheard of because the conventional wisdom is *"computers can't do anything new. They just do what humans put into them when programming."* This claim is already disproved when the computer understands a spoken text. What is understood is a new "emergent" quality, more than the acoustic signals of the voice. To be sure, humans have provided the method here. But in an evolutionary process, a solution is found that man did not or could not think of, and in the genetic algorithm, a method, a program, is itself invented.

Evolution is also for nature the method for something new, for emergence. Evolution has created or invented everything: the suitable chemical bases, the basic methods of life such as the functional solutions, the species. Chance receives its meaning in retrospect through life.

But even large technical software systems are themselves evolutions with a lot of randomness; in them we understand every detail and can thus learn the laws of evolution. The applicability of these Lehman's laws from the software industry to evolution is striking. Nature applies the principle of evolution and the same laws on a large scale, with very many participants, very many structures and feedback processes, and with much, even a great deal of chance. Some resulting phenomena can already be understood at the systems level. That was the luck for Darwin and his observations on the Galapagos Islands!

Evolution works with chance, which is confirmed or rejected *a posteriori*, lives on or perishes, and which thereby receives a direction. The structure of complexity is thus understandable except for critical points at which new megatrajectories are started, such as sexuality or, in particular, the first beginning of life in chemistry, abiogenesis.

No doubt chance is the driving force, but we cannot prove all the intermediate steps. It is in the nature of these beginnings that they are also difficult to verify. There remain (by their very nature) white spots on the map as a result.

The emergence of the complex lens eye has long been considered, on the whole, an insurmountable hurdle to evolutionary acceptance. It is psychologically tempting to see the complexity of the eye as a whole. Today, the formation is understood in great detail with many intermediate steps and variants, only the first beginnings, the formation of the required light-sensitive substances, are not.

Scientifically, philosophically as well as theologically, the question now remains: Is everything "ordinary" coincidence or are we something special, that is, a somehow "unique" coincidence in the cosmos. The sequence of concatenated positive events starts with the sun, then the earth, our blue planet, then the properties of water, the big moon, and it extends to the emergence of *H. sapiens* and even beyond that to artificial intelligence (considered as a positive development). Does the "Copernican principle" apply? That is, all this (that is, including us) in space is nothing special?

We can't answer the question, but we can show that simple answers like the "Intelligent Designer" are not entirely reprehensible, but they don't help and are not religiously advisable. It results only in an unworthy stopgap God.

Evolution is the method of nature to generate something new. We ourselves also belong to the generated. The machinery runs. Just like that. "It happens".

References

Blount, Zachary, et al. 2018. Contingency and determinism in evolution: Replaying life's tape. *Science* 362 (6415): eaam5979. https://doi.org/10.1126/science.aam5979.

Hehl, Walter. 2016. *Wechselwirkung – wie Prinzipien der Software die Philosophie verändern.* Heidelberg: Springer.

———. 2019. *Gott kontrovers*. Zürich: Vdf.

———. 2020. *Meine fünf Frauen*. Berlin: Epubli.

Kaufman, Gordon. 2001. *In the beginning . . . creativity*. Minneapolis: Fortress Press.

Kleine, Thorsten. 2019. *Formation of moon brought water to earth*. ScienceDaily.com. Zugegriffen am 21.05.2019.

Mack, Katie. 2020. Freeman Dyson's quest for eternal life. nytimes.com/2020/03/02/opinion/contributors/freeman-dyson.html. Zugegriffen im Juni 2020.

Madhavji, Nazim. 2011. *In memory of Meir Lehman*. pleiad.cl/iwpse-evol/keynote/slides.pdf. Zugegriffen im Juni 2020.

Nilsson, Dan-Erik, and Susanne Pelger. 1994. A pessimistic estimate of the time required for an eye to evolve. *Proceedings of the Royal Society of London* B525: 53–58.

Human Creativity and Chance

Creativity is historically a divine or semi-divine activity; as a field of psychological research, creativity has only existed since 1950. Creativity is closely linked to chance, visible or invisible. It is the way "the new comes into the world" (Klaus Mainzer). We analyze the process of creation of an idea or an invention. We thus follow the first analyses of the physicist Helmut Helmholtz and the mathematician Henri Poincaré, look at how an idea is created and define four phases: Preparation, brooding phase (the incubation), flash of inspiration (the illumination) and verification. All phases have chance in them, especially the first three.

> "Creativity was originally a term applied solely to the gods."
> Jon McCormack, Australian computer scientist.

In this sense, creativity was the creation of something new out of nothing, *creatio ex nihilo*. The creativity of the individual has replaced the gods in our understanding of creativity today.

7.1 Types of Human Creativity

7.1.1 When There Was no Creativity

> "The painter depicts the bedsteads that he sees realized in the material, so he is dealing with the after-image of the after-image of the table-self. And if the table-self is the truth, then the painter is far from the truth."
> Plato, Politeia 10th book, according to Martin Suhr, 2001.

© Springer Fachmedien Wiesbaden GmbH, part of Springer Nature 2021
W. Hehl, *Chance in Physics, Computer Science and Philosophy*, Die blaue Stunde der Informatik, https://doi.org/10.1007/978-3-658-35112-0_7

Fig. 7.1 Plato and Aristotle. Detail of the fresco in the Vatican. Plato on the left. Image: Plato and Aristotle in the School of Athens, Wikimedia Commons

In Plato's world (Fig. 7.1) there is hardly any room for creativity in the contemporary sense. On the one hand there is the real, the world of ideas (e.g. the idea of a bed as a bed for sleeping) and on the other hand the world of imitation (the carpenter turns a wooden bed according to this idea). The painter of a picture of the wooden bed is then only the imitator of the imitator. The idea is given, and the artist has to realize it "by all the rules of art." Imagination, for instance in the theatre, is not desired, but only distracts from the search for truth. Plato therefore banishes the poets from his ideal state.

Aristotle (Fig. 7.1) rejects abstract ideas that existed beyond perceptible things. He sees in everything new always already the existing basis or model. This applies, of course, to the constancy of animal species. An amusing problem for Aristotle is the mule, which he knows is barren (Jansen 2002). He obviously always has to arise anew. But the mule is not properly something new, it is a hybrid. The similar reasoning applies as in the explanation of the ancient proverb he cites: *"Libya is constantly hatching something new."* It is only hybrids that emerge when the animals crowd together at the few watering places (Hartung 2010).

We have already seen such simple combinatorial creativity in two recent examples, the combination disks of Ramon Llull and the combination matrix of Fritz Zwicky. In the modern understanding, these are primitive borderline cases of human creativity that practically lack the act of creation. The image of creativity is very limited: Man only recreates nature and makes illustrations to do so. And matter has always been there, even before the gods.

Aristotle reduces the divine to the unmoved first mover who lets everything go.

For Plato, in this understanding, all inventions are crafts. All artists are craftsmen who try to realize the given ideas. Even the world creator himself, the God who made the world, is for Plato a craftsman according to the ancient Greek word δημιουργός or *dēmi(o)urgós,* the "craftsman" or "builder." For Plato, ideas come from a fixed, divine store of ideas and blueprints; we can only work them out. This view is still present today in the word "creator" or "creation": we can only create from the quantity of what is available.

This world of ideas is opposed by the material world. So it is also a dualism like today with world 1, physics, and world 2', computer science, now starting with the complex blueprints.

Plato's attitude towards poets and their creativity is famous: he wants to expel poets from the state. However, not because they spread invented, mendacious things, but because they do not spread pedagogically positive things. The lie (the "fake news") is allowed for the good purpose of the state (Mecke 2015).

We are closer to the modern understanding of artistic creativity when Plato says that poets can only write poetry when they are drunk to unconsciousness (more generally, divinely inspired)—it is a foreshadowing of the idea of genius... Plato has Socrates say:

"Now the greatest of all goods are bestowed upon us by intoxication when it is bestowed as a divine gift. [... Therefore] intoxication, which comes from God, is nobler than prudence, which comes from men."

This also brings us closer to chaos and chance, and thus to the modern understanding of creativity!

In the Christian Middle Ages, the word *"creatio"* dominates from the Latin "I make", thought *"ex nihilo"*, out of nothing. The conception of creativity in this period was divided, the divine on the one hand, the human and merely imitative in art and technology on the other. A pioneer of the new conception was the Italian humanist author Giovanni Pietro Capriano (1520–1580); he wrote in 1555 in *"Della vera poetica"* (Weinberg 1961):

"The true poets must feign [fingere] their poetry out of nothing",
  'di nulla fingere la lor' poesia'.

Nothingness is close to chance in a modern sense, but he does not yet write *"creare"*—not yet the divine creating, but *"fingere"*—pretending. But the spirit of the artists of this time is in the dawn of independence and freedom in all spheres. Artists refer to their activity as *"thinking out"*, as *"predetermining"*, *"inventing what does not exist"* or *"condensing ideas"*. Leonardo da Vinci invents *"forms that nature does not know."* The architect Cesare Cesariano (1477–1543) describes the architect who realizes his ideas as a demigod, *"come semidei"*. The concept of the Renaissance man emerges, as embodied above all by Leonardo da Vinci.

While divinity thus passes to man, sober enlightenment develops at the same time. The mysterious act of artistic inspiration does not fit into this rational world, not even into the fine arts, as long as art is subject to rigid rules, as in antiquity. An example of this was the "academic style" in painting and sculpture from the 17th to the nineteenth century, which dissolved under the technical pressure of photography and the social pressure of a society that was becoming freer. The rules visibly disappeared, creativity gained the upper hand—but the word "creativity" did not yet exist.

The two paintings in Fig. 7.2 illustrate the transition from academic art to abstract art. The painting on the left is by the French academic painter Alfred Agache, member of the Société des Artistes Français, 1843–1915. On the right is a portrait of Picasso, painted by the Spanish painter Juan Gris, 1887–1927. On the left is the realistic immediate humanity, on the right the cubist, alienated and hacked figure, both persons inscrutable. On the left, chance is hidden in the details of the realistic scenery; on the right, it is roughly visible in the pattern of geometric forms.

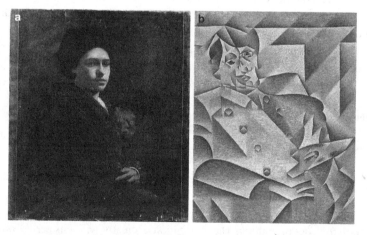

Fig. 7.2 Creativity in painting: chance subtle or evident. (**a**) "Unknown Woman," Alfred Agache, c. 1880. Image: Unknown Woman, Wikimedia Commons. (**b**) "Juan Gris paints Pablo Picasso," 1912. Image: Juan Gris Pablo Picasso, Wikimedia Commons

The word "creativity" appears surprisingly late for the ability to generate something new, starting in 1927 with the British philosopher Alfred North Whitehead (1861–1947). An example of his philosophical use of the term is the following quotation in his own translation:

> **"The creativity of the world is the throbbing emotion of the past hurling itself into a new transcendent fact. It is the flying dart of which Lucretius speaks, hurled beyond the bounds of the world."**
> **Alfred Whitehead in "Adventures of Ideas", 1933.**

The core of the flying arrow—as in the special case of evolution—is in our understanding chance, often hidden, but sometimes visible.

7.1.2   Forms of Creativity

> **"The final solutions to problems are rational, but the process of finding them is not." Joy Guilford, American psychologist.**

In our understanding, behind this sentence is chance, visible or invisible!

The beginning of creativity in today's psychological or psychologizing sense is a lecture by the American psychologist Joy Guilford (1897–1987) only in 1950. Guilford is mainly concerned with the phenomenon of intelligence and its analysis: intelligence is the sober brother of creativity. The concept of intelligence in the sense of psychometrics arose half a century earlier.

Serendipity
Chance is particularly conspicuous, indeed the crucial point, in the pretty stories of unexpected inventions with "luck-bringing chance," the serendipity, for instance, in:
the discovery of LSD, Penicillin, Post-It, Teflon, Velcro, the Viagra, etc. etc.

The original English word is preferably also used in German. The Dutch ophthalmologist Pek van Andel has collected and analyzed a thousand (!) examples of serendipities (van Andel 1994). One defines:

▶ **Definition**   A *serendipity* **is an unexpected discovery that leads to a lasting meaningful result.**

If at the time of the serendipitous discovery one was prepared for something different but similar, this is called *pseudo-serendipity*. In the list, for example, penicillin and Viagra are pseudo-serendipities, because Viagra was already tested as a drug, albeit for hypertension, and the penicillin discoverer was already an experienced antibacterial hunter and discoverer of the substance lysozyme. Sir Alexander Fleming wrote:

"You do not know what you will find, you may set out to find one thing and end up by discovering something entirely different."

The wonderful word *"serendipity"*, never to be forgotten, was invented in 1754 by the English writer and politician Horace Warpole, Earl of Orford, after a "silly tale" from Persia of the *"Three Princes Serendip"*; Serendip being the old Persian name for Sri Lanka. Serendipity is an established term; the contrary term *"Zemblanity"*, on the other hand, has remained a marginal term. The Scottish writer William Boyd (b. 1952) invented it for "unfortunate and anyway expected discoveries", in English "unpleasant unsurprises". The Siberian island of Novaya Zemlya stood godmother to this term—it was also the site of Soviet nuclear weapons testing.

A special kind of serendipity are good coincidences provoked by personal, special directed behavior. Neurologist and author James Austin has suggested the word *altamirage* for this, when one "forces" serendipity through persistent elaboration (Austin 2003). The term stems from the rediscovery of Altamira Cave in Spain and the painstaking discovery of the ceiling paintings after 10 years of cave visits. The helper for the discovery was the young daughter of amateur archaeologist María Justina de Sautuola, who suddenly called out in the semi-darkness, *"Papa, Papa! Mira! Toros pintados!"*—painted bulls! The quotation from Prosper Merimée, the author of the novella to the opera "Carmen", also fits the Altamirage coincidence, because a dog had "rediscovered" the cave:

"Perro que anda, hueso encuentra"—the dog that runs around finds the bone.

Or the statement of the inventor and philosopher Charles Kettering (1876–1958):

"Keep going and you will come across something, perhaps just when you least expect it." More academic, professional and famous is the saying of Louis Pasteur:

"Dans les champs de l'observation le hasard ne favorise que les esprits préparés"
- chance favors only the prepared mind.

It helps if you know as much as possible in your area and neighbouring areas!

There are obviously many types of serendipity, from raw randomness to elaborated randomness, which is again in a sense "inclined" randomness (see Propensity). We want to distinguish two types of creative creation altogether: New things with externally acting chance (serendipity) and new things by chance working internally, in the mind of the creative person during the creative process. Probably the first attempts to define such a process for inventions and scientific discoveries were made by the German physicist and inventor Hermann von Helmholtz (1821–1894) and similarly by the French mathematician and physicist Henri Poincaré (1854–1912). The linear structure consists of four phases:

1. Preparation and saturation: This is where the problem develops or where one explores all aspects of an already existing problem, e.g., prior knowledge, state-of-the-art.
2. Incubation (brooding phase): Distancing from the task and loosening up. In the background often doubts about oneself and the meaning of the task.
3. Illumination ("flash of insight"): A sudden insight into the context and the flaring up of a solution.
4. Verification and implementation: Possibly think through the idea the next day and plan the implementation.

After verification, the problem (with an unknown solution) becomes a task (with a given solution path) and the realization results in an innovation by definition. All four stages of work involve chance; stages 2) and 3), pondering and finding, are even dominated by chance.

The text in Fig. 7.3 gives an insight into the classical work of Henri Poincaré. In 1908 he wrote of the "*long travail inconscient antérieur, le rôle de ce travail inconscient*" and then of "*les apparences d'illumination subite*"—that is, of the sudden illumination after the long unconscious inner work.

Illumination can be a wonderful experience. Suddenly, out of the vaguely existing and changing patterns in the mind of the artist, inventor or scientist, a pattern emerges that "works" or at least promises to work. Henri Poincaré explicitly describes the role of the unconscious ("*inconscient*") and the subconscious ("*le moi subliminal*") in "*La Science et l' Hypothèse*" as early as 1902.

The mathematician Carl Friedrich Gauss described the following experience in 1805 in a letter to the amateur astronomer Wilhelm Olbers from the spirit of his time:

> **"Finally, a few days ago, I succeeded - but not to my laborious striving, but merely by the grace of God, I would like to say. Like lightning striking, the mystery has been solved; I myself would not be able to prove the guiding thread between what I knew before, that with which I made the last attempts, and that by which it succeeded."**

The anecdote of the chemist Auguste Kekulé (1829–1896) about the illumination of the discovery of the structural formula of benzene, C_6H_6, is famous. Indeed, the molecule of benzene has the form of a ring consisting of six carbon atoms. Kekulé wrote about his discovery as he had dreamed on the bus in London:

> **"I sank into reverie. The atoms were whirling before my eyes ... how everything was turning in a whirling round dance. I saw how larger ones formed a row and only at the ends of the chain still smaller ones dragged along ...**
> **The call of the conducteur, Clapham Road, roused me from my reverie."**

However, it could be just a pretty story, invented by Kekulé consciously as a punch line or marketing gag or unconsciously. We will introduce the latter possibility as "*press office*" in the decision-making process (Figs. 7.27 and 7.28).

Science et Méthode

INTRODUCTION

Je réunis ici diverses études qui se rapportent plus ou moins directement à des questions de méthodologie scientifique. La méthode scientifique consiste à observer et à expérimenter; si le savant disposait d'un temps infini, il n'y aurait qu'à lui dire : « Regardez et regardez bien »; mais, comme il n'a pas le temps de tout regarder et surtout de tout bien regarder, et qu'il vaut mieux ne pas regarder que de mal regarder, il est nécessaire qu'il fasse un choix. La première question est donc de savoir comment il doit faire ce choix. Cette question se pose au physicien comme à l'historien ; elle se pose également au mathématicien, et les principes qui doivent les guider les uns et les autres ne sont pas sans analogie. Le savant s'y conforme instinctivement, et on peut, en réfléchissant sur ces principes, présager ce que peut être l'avenir des mathématiques.

Fig. 7.3 The first page of "Science and Method" by the physicist Henri de Poincaré from 1908. The text exemplifies ideas of our model for creativity. Image: Document numérisé de la Bibliothèque Interuniversitaire Scientifique Jussieu UPMC

To this day, dozens, probably hundreds of methods with courses and books entwine around these processes for creativity and management in general, for the management of innovations or the creation of works of art in particular. For the "incubation" stage in particular, there is advice such as *"don't learn too much of what exists"* (so as not to block yourself), *"think outside the box"* (to come up with something completely new) and drink alcohol in moderation (to loosen up mentally).

Here's a title from Psychology Today, April 4, 2012:

Alcohol Benefits the Creative Process
Being moderately intoxicated gets people to think "outside the box".

On the other hand, there is also a correlation between creativity and mental illnesses, e.g. borderline disorders!

The word "chance" (or "hazard" or "Zufall") hardly appears in the literature of creativity, but the first three stages of the process are full of chance: the selection of areas, the individual comparisons, the memories that come up, and of course the flash of inspiration itself, or at least the preconditions for it, are random. As a synonym, the psychologist Joy Guilford uses the term "divergent thinking" for thinking determined by chance.

Level 4 means "convergent" thinking, it is the area where "creativity" turns into "intelligence". We define both terms:

▶ **Definition** **Creativity is the ability to create or think meaningful new things. The context determines what is meaningful.**

In 1790, the German classical philosopher Immanuel Kant speaks of "original and exemplary" in connection with the creative artist, i.e. the new must endure and continue in the respective field. For the ability "intelligence" we define:

▶ **Definition** **Intelligence is the ability to solve a complex task, even if there are still uncertainties in the description or the solution path.**

The uncertainty consists in the ambiguity of the boundary conditions and initial values, in the size of the problem or in the indeterminacy of the solution method. If everything is clear in the problem, in the problem definition as well as in the solution method, and the complexity consists only in the large numerical scale, then the solution does not actually require intelligence. A "calculator" with a deterministic algorithm solves the problem with "brute force", with brute computational force and obedient following of the instructions. Randomness would only be disruptive. Even in large-scale tasks, it is the trivial limiting case of intelligence: the complexity of the solution is in the algorithm. This corresponds precisely to the distinction between "problem" and "task".

- With creativity, chance is at the center.
- Intelligence is about the correct handling of chance, namely the sensible and successful incorporation of chance into a rational solution system.
- With the algorithm, you want as little chance as possible, because here chance would be an error.

The three stages from determinism to indeterminism, much chance, little chance to no chance at all, we illustrate with the example of board games, such as Mill or Checkers or Chess or Go:

(a) *"Creativity"*: without much knowledge of the game or when there are a lot of possible moves.

At first, chance dominates. Methods develop to better deal with chance.

(b) *"Intelligence"*: with knowledge and experience, but many opportunities develop patterns of experience and strategies.

We define strategy as a small system of rules that allows action without having a full view of a situation and in the expectation of chance.

(c) *"Algorithm"*: the game is transparent in all possibilities, either every move can be calculated through or there is a library with all possible moves.

An algorithm is not a game, it is a working instruction.

In the course of the temporal development of the power of computers and the corresponding game software, games pass through the phases (a) to (c), just as a human being passes through the phases beginner, good player and (eventually) "champion". To the beginner, many things appear to be coincidental; to the expert, they appear to be lawful. When a player or software sees through the game in all its possibilities, then everything is just an algorithm. The algorithm is by definition a clear course of action that leads to the goal.

For example, in the game of mills, the entire space of possible positions can be stored in a database of 17 gigabytes for a perfectly playing program, and in the game of checkers, correspondingly, all 10-man endings are in a database. Wikipedia says laconically:

"Today, [checkers] programs running on PCs can't actually lose to human opponents."

After humans were overtaken by computers in chess by IBM's *Deep Blue* in 1997, this happened two decades later in the game GO, which is considered the most complex abstract game. Since 2015, variants of Google's *AlphaGo* software have been winning against the best human players. Figure 7.4 shows a small-scale variant of GO on a PC with aids and on a 9×9 playing field instead of 19×19.

The progress of information technology is constantly removing chance from these games and problems and replacing it with knowledge. When chance and uncertainty disappear altogether, this is no longer a real feat of intelligence by our definition—nor a game, but work.

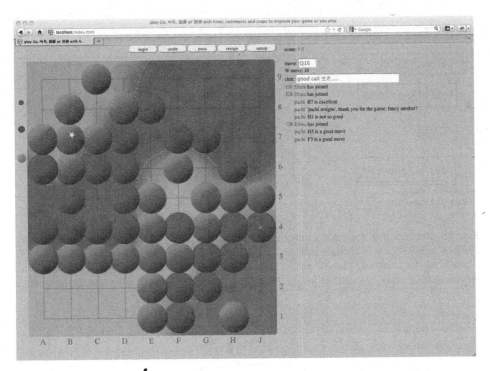

Fig. 7.4 The game GO in a 9 × 9 version on the PC. With aids. The game GO is considered one of the most complex games. Image: 9 by 9 Go Game with Maps, Wikimedia Commons, Signbrowser

Creation by Association

In the examples of serendipity, the intrusion of chance is wonderfully visible, indeed it is what makes the stories so appealing. To understand, or at least illustrate, a creation in the mind, we use the vocabulary of *external* creation by Pek van Andel, who collected the design patterns for the visible emergence of creations using the example of serendipities.

The diagram in Fig. 7.5 is intended to remind us of the philosopher Llull's wheel in Fig. 6.4: Chance establishes the most diverse connections of thought patterns with the fields of knowledge and checks their meaningfulness. The scheme of Fig. 7.5 is associative, i.e. associations to the task are (randomly) sought and analyzed. The figure applies to two types of creations, namely, on the one hand, to concrete, given tasks and, on the other hand, to free creations whose possible meaning is determined only after they have been successfully produced. For the task "free artistic creation" we will still present below the model of coupled associations, the "bisociations" according to Koestler (Fig. 7.6).

For simple examples the matrix of Mr. Zwicky, which we already got to know, is sufficient. A more complicated method for engineers and with a larger variant matrix is the TRIZ matrix method for systematic inventing, which originated in the Soviet Union.

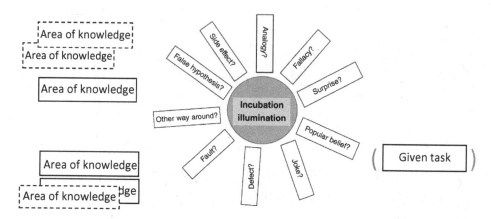

Fig. 7.5 Illustration of the emergence of a creation by association. Directed chance continuously connects knowledge with thought patterns. In the circle a selection of creative patterns, on the left symbolically the knowledge areas ("mental spaces") that are examined and on the right (optional) the specification of a task, if available. Image: The thinking patterns according to van Andel, image own

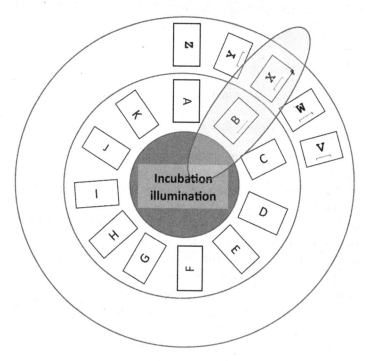

Fig. 7.6 Illustration of the emergence of a creation with bisociation or "conceptual blending". In the sense of Ramon Llull, the two circles symbolize different conceptual spaces ("mental spaces"), which combine in different ways when twisted

In the mind of the inventor or problem solver, "directed chance" is at work again: some areas of knowledge are closer than others, some test question is more important than another.

There is chance involved because the number of elements stored is too large and access is fuzzy. If the solution to a task is clear—such as "it's dark and there's a flashlight" or "there's a banana and there's a stick to fetch it"—then little chance is involved. But then it's not a creative effort (for humans, at least)—but it might be for chimpanzees.

At the exit of the illumination, the idea is tested. In the case of mathematics, it is calculated on the blackboard; in the case of a possible engineering invention, it is drawn; in the case of software, a prototype is designed; and in the case of a work of art, a draft is sketched. A painful experience can be that the verification of such a good idea turns out to be sobering, as the author knows from his own experience: The invention doesn't work after all!

In the teachings and therapies of the Swiss psychologist and psychiatrist Carl Gustav Jung (1875–1961) these phases of incubation and illumination play a great general role; here he uses expressions from late medieval alchemy: *imaginatio fantastica* is the world of free ideas and patterns, *imaginatio vera* the realistic one: the solutions that really work. It is where the theory becomes reality and the ideas are brought *down-to-earth.*

Creation and Bisociation: The Mixture of Concepts

> "The humorist, on the other hand, deliberately chooses discordant codes of behaviour or universes of discourse to expose their hidden incongruities in the resulting clash. Scientific discovery may . . . be similarly described as the permanent fusion of forms of thought previously thought incompatible."
>
> Arthur Koestler, Hungarian-British writer, from "The Act of Creation," 1964/1989.

In contrast to association, Arthur Köstler (Koestler 1964) sees the new arising from the combination of opposites. His prime example is the discovery that the fields of electricity and magnetism belong together and together explain light, for example. He coined the term *bisociation* for this; the psychologists Gilles Fauconnier and Mark Turner refer to it as "conceptual *blending*", the merging of conceptual spaces in the subconscious (Fauconnier and Turner 2002).

The two rings of Fig. 7.6 are modelled on the discs of the Mallorcan philosopher Ramon Llull: Twisting the rings brings two conceptual spaces together into pairs, in the image X and B. In humans, chance acts invisibly here. The synthesis of X and B can be attempted in the subconscious. Maybe the person carries a thought from space X for months and unconsciously hundreds of ideas from the other conceptual spaces are compared. However, nothing happens in the subconscious as mechanically as the rotation of the disks in the analogue—it is again "directed chance" in the sense of Karl Popper.

The Genius and the Coincidence

"There is no great genius without some touch of madness."
Aristotle, Greek natural philosopher, 384 B.C.-322 B.C.

"If people knew how much work went into it, they wouldn't call it genius."
Michelangelo Buonarotti, Italian painter and sculptor, 1475–1564.

The concept of genius had its heyday during the Romantic period. It came into the German and English languages via the French *génie* from the Latin *genius*, originally "the generating force". Genius exists (or existed) in the most diverse fields—scientific, mathematical, political, economic, but above all in art. In Romanticism, the genius united in itself the Romantic attitude to life. Almost supernatural powers were attributed to him, e.g. to see more and feel more than a normal person. Today, genius is spoken of more journalistically, less in the sciences.

For "geniuses" and their works of genius, there was the presumption of something special, perhaps even outside our rational explanation with chance lying. Then the genius would be outside the possibility of artificial intelligence. One of the assumptions was an extraordinarily high general intelligence of the genius—but this does not seem to be the case. Rather, individual cognitive traits are preeminent in the subject area. The romantic ideal of the genius is mainly matched by extreme creativity and high productivity—albeit extreme to the point of bordering on the psychopathological fringe, even to the point of preliminary stages of schizophrenia.

One reason for the increased creativity is probably the characteristic of lower "*hidden inhibition*" (latent inhibition) in the so-called geniuses:

It is possible to see a stimulus more often without it being particularly noticeable, indeed to get used to it. The chance to perceive it consciously decreases continuously. The stimulus is to be seen (or the possible association is obvious or "actually the thought is obvious"), but is not perceived by this "latent inhibition"—it is a protection against overstraining. This inhibition threshold is lower in the case of genius, at least in the respective discipline of these people. If factors such as a community of admirers, good marketing with exceptional attributes, e.g. in appearance or behaviour, the right zeitgeist and a lucky coincidence are added, the career to genius can take place.

The genius does not possess supernatural powers, but creativity (and chance) are pushed to the extreme. This means that the computer can also be a genius.

7.2 Creativity and Computers

7.2.1 Computers Are Creative with Chance

"Les ordinateurs sont inutiles. Ils nous donnent que les réponses."
Computers are useless. They only give us the answers.
Pablo Picasso, Spanish painter and sculptor, 1881-1973.

The interpretation of the quotation would depend on the time when Picasso said this—but I could not fix it more precisely than "before 1973". But the content several times misses the facts, which were, of course, difficult to see at the time.

The most famous answer of a computer is probably "42". It is the answer of the computer Deep Thought to the rather indeterminate question "about life, the universe and all the rest". This indeterminacy is the problem, see Wikipedia article 42 (answer).

More precise and also more general is to understand the computer as a command receiver (but this does not look more creative at first): If you press a key like "remove" it actually means "Remove!", with "copy" it is the command form "Copy!". But there are already executable higher, more "creative" commands like "Format!" (put a text into a nice, clean shape) or "Translate!" (e.g., translate a spoken sentence from Mandarin to German). Since about 60 years there is even the factual answer to the command: *"Compose and play a piece of music in the style of Bach"*, *"Density"*, *"Paint"*, and today even *"Build in 3D"* with the help of the 3D printer.

From the mid-fifties of the last century, the circle around the philosopher and physicist Max Bense (1910–1990) conducted the first experiments with "art and cybernetics" at the Universität Stuttgart, then the Technische Hochschule: poems and graphics were generated on the Z22 tube computer and output with a punch card and plotter (a computer-controlled drawing device with pens). It is astonishing how far one can already get with the simplest means, i.e. what is recognized by the human audience as a human product or as an apparently human product. We have already mentioned above Joseph Weizenbaum and the successful story of his minimal-intelligent program *Eliza*.

A small example of this was a simple but successful type of puppet called IBM Pong, around 2000. Actually just an oval disc with two moving eyeballs that could follow the observer—but the puppet still looked lifeless. If you added a permanent small random trembling movement to the eye movement, it seemed alive. It is something like the clinamen of the atomists 2200 years ago! The coincidence is in more than one sense necessary for life (or to appear alive).

Max Bense wrote *"Chance is given, opportunity is seized."* This is a good description of the function of chance! Creativity comes from the sum of rules and the injection of chance, which serves as input to provide opportunities for various actions. At its simplest, randomness is simply a "truly" random sequence of numbers read off or generated on the fly. In practice, one often uses a sequence of near or "fake" random numbers. Each computer-generated work is created with new random numbers and is again an original.

Today's programs combine finding the rules with learning from many examples. It was clear from the beginning that music is particularly well suited for working on and in the computer; after all, sheet music is actually also software:

The process of composing is a kind of programming, the notes are the program and the performing musician is a kind of computer-controlled output device. In composing, the composer creates and observes chance itself in his head.

Fig. 7.7 The chorale *Wenn wir in hoechsten Noethen* from *the Art of Fugue* by Johann Sebastian Bach. Excerpt from the first printing. Image: Bach Art of the Fugue first edition, Wikipedia.de, Rarus

We remember: In the world model, this corresponds to computer science actions in world 2′; only in the optional world 3′ does music possibly become something more, namely true art.

The beginning of artificially composed music was already made in the fifties of the last century with the first music created by the computer. The music theorist and chemist Lejaren Hiller already created the *"Illiac Suite"* in 1957, a string quartet with the Illiac brand computer.

A great example in music are the chorales of Johann Sebastian Bach (Fig. 7.7). They are great to hear as music, great in the intricate sequences of notes, and they are great targets for artificial composition in a given style: they bring together melody and harmony according to strict rules, but also contain (or need) sufficient chance to be alive. A modern program is DeepBach by Sony Labs Paris (Hadjeres et al. 2017). From the 389 known Bach chorales, it has extracted the style with "deep learning", by learning all the compositional patterns, and *now "can do Bach"* at the push of a button. This music is indistinguishable from the "old" Bach for musical laymen and difficult for professional musicians.

The visual arts have undergone and are undergoing a similar development since the beginning with computer-controlled plotters from high-resolution videos to 3D printers. As a reminder: we command the computer to compose, draw, make music or even build. The computer, for its part, translates the task into notes, pen movements or drops of ink or blobs of matter in the right place.

One of the first and most famous realizations of visual art in the computer was the Aaron system by the British artist and computer scientist Harold Cohen (1928–2016). Aaron was particularly capable of producing representational images, such as still lifes or portraits. Figure 7.8 shows vases and plants in color; the image was made in 1995 and is in the Computer Museum Boston. Each new skill of Aaron the artist had to be painstakingly professionally programmed. In this sense, it is a joint work of the human Harold and the program Aaron and, of course, chance. In our definition, both man and computer are

Fig. 7.8 Image of Harold Cohen produced with the Aaron program system, 1995. From the Computer History Museum Boston, blog post 40 years Harold Cohen and Aaron

creative, the machine somewhat more so than the man, who is, after all, essentially implementing known rules.

Courtesy Computer Museum Boston and Harold Aaron Trust.

Cohen himself is much more cautious, saying:

> **"If what Aaron is making is not art, what is it exactly, and how is it different from the 'real thing'? If the computer doesn't think, what exactly is it actually doing?"**

With our definitions, it is technically clear that his computer DEC VAX 750 thinks. The creative task is divided between the digital computer and the human computer in the artist's head. It is more difficult to assess the value of the objects created.

Several human aspects make value problematic:

- The advantage of being able to produce any number of different works of almost any complexity at any time is problematic under the laws of the market. Uniqueness would raise the price. The American computer composer David Cope (b. 1941) is said to have once left his EMI (Experiments in Music Intelligence) program running by mistake over lunch. When Cope returned, the computer had composed 5000 "original" Bach chorales.

  In the case of copyable works, the artificially limited edition suggests itself as a practical solution.
- The advantage of being something new wears off very quickly.

AI Art at Christie's Sells for $ 432,500

Oct 25, 2018.

Last Friday, a portrait produced by artificial intelligence was hanging at Christie's New York opposite an Andy Warhol print and beside a bronze work by Roy Lichtenstein. On Thursday, it sold for well over double the price realized by both those pieces combined.

Fig. 7.9 Image generated with Artificial Intelligence and next to it the sales note in the New York Times. *"Portrait of Edmond de Belamy"*. The image here is in the public domain because it was created by the artificial intelligence of the French collective Obvious. (Source: Edmond de Belamy, Wikimedia Commons)

On the other hand, especially in the field of IT and digitalization, there are constantly new developments, also of a fundamental nature, and an almost inexhaustible reservoir of experimental talent.

- The abstractness of a software or machine is a disadvantage on the market. The wisdom of the pre-Socratic philosopher Protagoras (about 490–420 BC), the so-called Homo Mensura theorem, explicitly applies here:

  "The measure of all things is man, of those that are, that they are; of those that are not, that they are not."

  Above all, what counts for us humans is the value attached to a thing by other people.

Figure 7.9 from a Christies auction shows that—despite the problems above—works of computer art are a growing market and, as is common in the art market, are risky investments, just like other modern directions of man-made art. The computer-altered portrait *"Edmond de Belamy"* shown and sold was obviously on an art market par with Andy Warhole and Roy Lichtenstein, the two most famous exponents of Pop Art, on October 25, 2018.

Undeniably, the computer as a craftsman has the advantage of being indefatigable and of giving the art object any number of details, thousands or even millions, scooped out of millions of coincidences. In the visual arts, this first brings to mind the man-made computer-like works of Victor Vasarely and M. C. Escher. When Vasarely writes (Smale 2005):

*"The units of my works: circles and squares in many colors, correspond to stars, atoms, cells
and molecules, but also to grains of sand, stones, leaves and flowers"*

This is even more true for the computer, even three-dimensional. The details can be three-dimensional and not even possible for humans in detail. In architecture, the computer designs and builds unimagined new constructions, in film entirely new worlds.

So computer art can be art according to the logic of the market. Art is actually a possible part of world $3'$ in our world model. Perhaps art defined via the market is at least a kind of shadow world to it!

7.2.2 Computers Think Almost Humanly, Even without Understanding

"Deep neural networks are responsible for some of the greatest advances modern computing has made".
Jeff Dean, American computer scientist, born 1968.

Technically more sophisticated and philosophically more interesting are the works of the computer, which are created without detailed technical specifications. A basically simple basic technology for this are neuronal networks, which can learn something Without understanding it—just like we do. We have already mentioned John von Neumann's remark about understanding in mathematics.

It also depends on the respective definition of "understanding":

▶ **Definition Understanding a process can mean having a model of the process that goes at least one level deeper than is obvious, e.g., "I understand that XYZ stock went up today because XYZ company announced a great new product."**

This definition is deliberately not absolute, but only relative. If the understanding at one level is not sufficient, one researches further ("deeper") in science or, in technology, calls in the "deeper" expert for a problem, or, as a last resort, the developer himself.

A running neural network is like a filter glass through which one sees the world, i.e. receives it, or through which, conversely, one sends an image, for example by sending a noise image through it. By sending very many (possibly millions of) cat images, the filter "cat" is created; with this filter, cats can be found in images or, conversely, cat images can be created or inserted into scenes. Such a filter creates a worldview with a propensity to see cats, it is a kind of propensity according to Popper, but now not physically the propensity of a snowflake to find hexagonal points, but in the living to see or project the programmed objects. The computer learns with the given examples all by itself, what a cat is, also it could learn by itself, that these objects probably are called *"Katze"* in German, in English *"cat"*. Computers learn, respectively the software learns, although a popular wisdom says:

Fig. 7.10 Image generated with Google DeepDream after template "three men". Image: DeepDreamScope, Wikimedia Commons, Jessica Mullen

"Computers *cannot learn anything*". Philosophically speaking, the program develops the Platonic idea *cat* from the Aristotelian externality.

One example is Google DeepDream, a neural network that initially recognizes objects, but primarily inserts structures into given images under internal randomness (Mordvintsev et al. 2015). It is a "deep" neural network, meaning that multiple networks or filters are connected in series. The output of one network is the input to the next level at about 10–30 levels above. So the first layer first sees corners and edges. A middle layer already sees single objects like a door or a leaf, the last layer then constructs holistic things, a building or a tree.

To a real picture the ghostly additions are added by the program with the learned structures (Fig. 7.10). The press calls the results and the impressions from "simple" to "artistic" to "horrible" or from "beautiful" to "scary".

Indeed, psychedelic worlds of Jungian manner emerge, which could have come from Roten Buch of CG Jung. Thus, Fig. 7.10 shows three male figures whose inner representation has been greatly intensified in the system to the point of giving the impression of archetypal, human, Middle American forms.

Videos of "deep diving" into the world of the lower layers are particularly impressive. A "deep zoom" technique is generally an almost spiritual and philosophical experience through the emergence of ever new random or coincidental structures.

Here are three types of deep zooms:

Fig. 7.11 DeepDream image created from noise alone. Image: Deep dream white noise 0028, Wikimedia Commons, Martin Thoma

- Zooming deep into the real world: from atoms to galaxies.
- Deep zooming in the mathematical world, namely into the Mandelbrot set.
- Deep zooming into the psychedelic world of DeepDream.

Zooming into the real world from atoms to clusters of galaxies is about 35–40 orders of magnitude of physical structures and randomness, sinking into the exact but pseudo-random looking world of the Mandelbrot set is about more than 100 orders of magnitude, spinning into the psychedelic depths is about indefinitely many orders of guided randomness. All these journeys into depths make us shudder, physics and mathematics show us our smallness, with each step into the larger or smaller more, and the zooms into the inner soul (or at least the inner software) become more and more mystical and threatening.

Creation of the image from "nothing". Image: DeepDream White Noise, Wikimedia Commons, Martin Thoma.

The highest level of creativity is the growth of realistic or slightly to strongly psyche-delic images out of a pure, colorful noise without consciously given information—just as dreams partly arise from what one has seen and experienced the day before (Fig. 7.11). In the picture, dogs or "dog-like" grow here.

We can see this imaging technology philosophically as propensity made visible, visible propensity of randomness in a system. With software like DeepDream, randomness is influenced by imprinted, unconscious psychological patterns corresponding to physical randomness through the symmetry of water molecules.

Here are some stages of "processing" an image, ranging from unintelligent (physics only), to automatically changing (via electrical engineering), to intelligent manipulation of the input (with software):

- Droste effect: Optical image of a scene into the scene, such as multiple reflections. To be experienced between two mirrors; one looks into the abyss or falls into the depths. This is also called *mise en abysme,* English "pushed into the abyss" or Droste effect after a cocoa can that depicted itself on the package. The alienation does not change anything about the object, it is only "an affine image". We humans identify the object anyway.
- Video feedback: Pointing a camera at a monitor that shows its captured image creates a feedback loop that is either influenced by a co-recorded object or arises from the initial noise alone.

  Figure 7.12 shows the response of the system when the camera is rotated 45° and looks at the corresponding monitor. A view into the depths, into the abyss, framed by squares and octagons, is created on the monitor.
- Deep neural networks (Google DeepDream): generation of new textures and new figures, e.g. of fish or birds, as previously learned.

The images in Fig. 7.13 show a section of the view from the balcony of Fig. 3.14 in two versions. They are slightly psychedelic enhancements of the same part of the image, called up twice with the same program and the same settings. The DeepDream program obviously knows fish and birds very well—different specimens are visible in the image,

Fig. 7.12 Simple video feedback with twisted camera. One is also *"mis en abysme"* by this. Image: Video Feedback Octagon, Wikimedia Commons, DeathGleaner

Fig. 7.13 Detail of the original image of Fig. 3.14 (balcony scene with greenery and lake). The same place processed in two runs in a row by DeepDream with the same settings. (Source: own image, twice DeepDream in Deep Dream mode, Inception depth normal)

in the left image a fantasy animal is sitting on the balcony railing, on the right it is exotic birds.

The strongest random (or pseudo-random) changes give the "deep dreaming" of the software. Here one can even see disturbing propensity. These scenes and videos show in image form which impressions are dominant in each layer of the multi-layered creative network; the developers call this "inception" after the film of the same name, in which one can penetrate the dreams of other people, get information there, but also deposit information.

It would be interesting to see the work of a truly three-dimensional version with moving, growing and disappearing monsters—captured with instantaneous prints from the 3D printer.

We consider the Inception as a teaching piece and as a visualization for the philosophical qualia problem: Here one sees namely how sensual impressions arise in another "living being". On the computer example one can follow it, one can stop the program and watch on the different levels. On the lower level you see special textures, on the higher level shadowy animals. Of course, even for the same program, the process is always somewhat different—other coincidences, other preconditions, other scenes.

With ourselves, with our flesh and blood brain-computer, it is more difficult to see the inner workings, especially because we are "in it" ourselves. We are the computer. The idea of lower layers, or unconscious layers at all, "below" our consciousness, has arduously emerged. It comes from philosophy. For example, the German philosopher Immanuel Kant says around 1833:

> "... *that the field of our sense-perceptions and sensations, of which we are not conscious, is immeasurable, the clear, on the other hand, receive only infinitely few points of the same, which are open to consciousness.*"

The author of the DeepDream program wrote at the start of his software:

> **"We wonder if neural networks could become a tool for artists ... or perhaps shed light on how creativity works more generally."**
> **Alexander Mordvintsev, Google Designer, 2015.**

Both can probably be affirmed at least partially in the meantime.

7.3   The Human Being in the Computer Model with Randomness

7.3.1 Human Creativity and the Computer Model

> **"They're made of meat!" "Meat?". "Yes, made entirely of meat. We have taken some apart. They are meat through and through." "That's impossible. Who made the signals?" "Machines. They made machines." "That's ridiculous! How can meat make machines? You want me to believe in sentient meat? I am not asking you, I am telling you."**
> **Terry Bisson, American author, in "Meat", 1990.**

The Brain as Electrical Machinery

The lead is from the famous and also filmed short story "They are made out of meat"— 'They are made of meat'. Wikipedia says in the article brain (food) to the brain from a material point of view:

> "*Brain is made up of a soft, grayish-white mass composed mostly of roughly equal parts fat and protein.*"

Already the basis of our mind is unimaginable from a naive point of view. How can electricity arise in the brain at all? The incredible idea of electricity in the flesh got its start on November 6, 1780, when the Italian physician Luigi Galvani discovered the twitching of frogs' legs under the influence of electricity, and a few years later the Italian physicist Giovanni Aldini made their faces twitch during galvanic experiments with the heads of executed people (Fig. 7.14).

Research into the physics of the brain began with the work, or rather thoughts, of the Danish psychologist Alfred Lehmann (1858–1921) on the conservation of energy in the

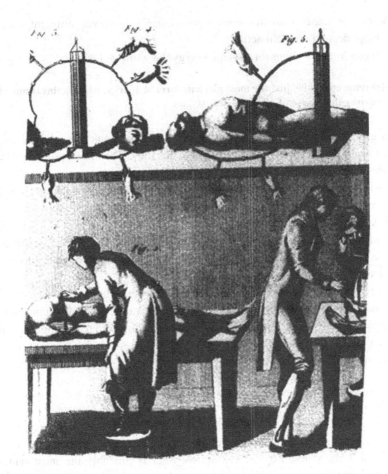

Fig. 7.14 Electrical experiments by Giovanni Aldini on the executed double murderer George Foster. Image: George Foster, Wikimedia Commons, Lokilech

brain system. The theorem of conservation of energy was striking—couldn't there even be a psychic energy in the brain?

The next milestones on the way to understanding the technical functioning of the brain were the discovery of weak currents on the living animal body by the English physician Richard Caton around 1875 and the first measurements on humans by the German neurologist Hans Berger around 1924. He created electroencephalography, a method of recording voltages measured externally on the scalp. Our head emits electrical (and of course magnetic) noise to the outside at frequencies between 0.1 and 30 Hz[1] and with an

---

[1] Typical frequencies of modern computer chips are 2 to 3 GHz.

amplitude that decreases with frequency ("pink noise"). However, different patterns are embedded that depend on brain activity.

Hans Berger was searching for psychic energy; he wrote in his diary:

"P.E. [psychic energy] is just the most glorious form of energy, which gains tremendous influence over the course of all processes ...".

For them exactly also the law of conservation of energy should be valid, as there is a mechanics-heat-equivalent! P.E. does not exist in this form. The mental processes consume energy, about one third of our total energy, for switching neurons.

All electrical and electronic devices, including computers, emit such electromagnetic radiation because currents flow and are switched in them. However, the developers of a technical device must take care to be "electromagnetically compatible". This means they must not interfere with other devices (or themselves). Interfering with itself means that parts of the electronics still introduce sufficiently strong signals into other, separate parts and then interfere with their function.

In the brain-computer interface (BCI), computer analysis of brain signals is used to send signals from inside to outside the brain. This enables a paralysed patient to give commands to a wheelchair, for example. Technically, it is a matter of pattern recognition, the filtering out of meaning from the multiple noise in the typically 32 measuring electrodes.

The dream of the pioneers Lehmann and Berger was:

"A real psychoscope, an apparatus by means of which one can diagnose with no small degree of certainty the state of mind of a person." (Lehmann and Bendixen 1899).

Today, there are a number of neuroimaging techniques for such "psychoscopy" of normal and abnormal brain states. Figure 7.15 shows what is probably the most widespread "functional magnetic resonance" procedure, which essentially shows areas of increased blood flow due to increased activity. This reminds the historian of Aristotle: for him, the brain was an organ for cooling the blood ...

The procedures show that whole waves of electrical excitation run through the brain. The brain is a system full of possible coincidences:

- On the one hand, it is the large number of elements involved, many thousands of nerve cells (neurons) running synchronously in a wave.
- In total, the human brain has 86 billion cells, 16 billion of which are in the cerebral cortex.
- In addition, the connections from cell to cell are the synapses, about 100 trillion of them, per neuron about 1 to 200,000 connections.
- Each neuron receives spikes, electrical or chemical pulses from thousands of neighbors, and these from others, and so on.

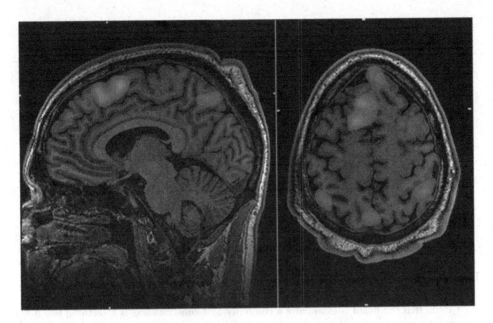

Fig. 7.15 Image of an fMRI scan. Zones of brain activity are visible. Image: fMRI scan during working memory tasks, Wikimedia Commons, John Graner

- The lowest level of randomness or directed randomness is the firing of a neuron itself: Neurons fire constantly, about 0.1 to 2 times per second, and up to 1000 times per second when stimulated.

In short, the brain is an impressive, massive, functioning stochastic system. There are anatomically fixed connections, but the running system "brain" is dynamic, flexible and directed—random. It is amazing that you can read this text accurately despite all the stochasticity!

Figure 7.16 illustrates the recognition of an apple as a small example of multilevel pattern recognition intended to show the similarity of human recognition and computer recognition. Level (a) shows the apple mixed with leaves as neutral objects. But the two types of data are processed differently in level (b): the apple as food, the leaves as "uninteresting". From level (c) and higher, the object is further interpreted, for example classified according to the apple variety, compared with the favorite apple, or even higher considered as a cultural object. This analysis is realized by *adhoc* closed and strengthened connections and weights of the neurons, very similar to what we have seen in artificial neural networks for example in DeepDream. We are also a large network of neural networks.

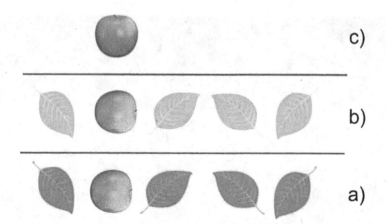

Fig. 7.16 Levels of attention becoming connectivity. Image: own

The Brain as a Computer Analogy

> **"I think that the brain is essentially a computer and consciousness is something like a computer program. It will stop when the computer is turned off. Theoretically, it could be recreated in a neural network, but that would be quite difficult because you'd need all the person's memories."**
>
> **Stephen Hawking, British physicist, on the question: What do you think happens to our consciousness after death? Times, 2010.**

This saying is quite elementary, but probably expresses what the brain and all living things are in general: they are objects of world 2′, the world with accumulated blueprints. Accepting the difference between world 1 and world 2′ was not only historically difficult. Many people today are still *du Bois'*, as described above: How is 'red' created? what is a 'thought'? what is 'knowledge'? what is 'identity'? what is 'death'?—these are questions about objects of world 2′. One can only understand them if one accepts Hawking's picture.

Historically, there have been three classes of explanations of the mind (or attempts to do so)

1. Physics (world 1): "Brain cools the blood", a Frankenstein electric machine, . . .
2. Metaphysical: "Somehow", "supernatural", such as a "channel system for spirit" (Descartes), . . .
3. Computer science (world 2′): Complex computer science based on physical chemistry. No more.

In principle, the mind cannot be understood from physics alone. As a constructed world 2′-object many things, at least on a philosophical level, clarify themselves abruptly. The mystical hypothesis is literature or psychology. The adherence to mysticism is more emotional: it is important for us humans to be special. This, of course, applies to knowledge and language (and driving).

But both in the human body and in our human information technology, everything is natural and thus even replicable (see below).

Some clarifying thoughts are repeated for their importance:

- Computing doesn't have to be in a box, doesn't have to run on transistors, doesn't have to be digital,
- Interactions of a hardware (executing function) and associated software (commanding and function) exist in many pairings,
- Complex hardware like a computer chip or the brain are actually also software, because they are built according to a blueprint: The chip with little, the brain with much chance or 'Zufall'!

Computer chips are also developed on the computer with software—in this sense, software is the central concept of the world 2' and in software its complexity is the essence (Hehl 2016).

Our man-made software systems are already gigantic as well; individual programs easily span 10 to 100 million lines or, put another way, thousands of man-years of systematic work, not quite as haphazard as the work of evolution in its four billion years. There are laws (or recommendations) about how to build large software systems, and our assumption here is: *These laws also apply to the construction of the big brain, e.g.*[2]

- Large systems are built in layers that are more tightly coupled within themselves than between them.
- You summarize tasks and determine what you're going to give them, what you expect them to do.
- You continue to use the tried and tested.

And more special:

- The neural network technique is useful in humans as well as computers.

The "laws" of building large software systems have been painstakingly learned by the computer industry, for example in the construction of communicating computer networks in the 1980s. Large systems must be built systematically to manage complexity. The author experienced the problems of historical all-in-one programs himself: They were no longer manageable.

From the software engineering point of view, we distinguish three layers (Fig. 7.17), which also correspond to three areas of responsibility.

---

[2]Lehman's laws for software systems, mentioned earlier, concern the life of the systems, not the initial development.

Fig. 7.17 An elementary model
of the brain with three levels in
terms of software engineering

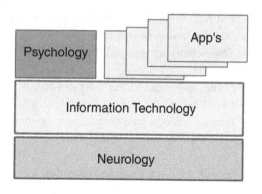

The basis is the circuits and processes on the physical and chemical level, which form the basic neurological elements and provide basic functions such as time and space perception and highlighting of changes. The middle level is usual informatics with pattern recognition and neural networks, storage and addressing processes and memories with access speed, scope and storage duration. The main IT infrastructure task is consciousness, which we use to manage our practical lives (see the definition below). At the level above that is the psyche, our most important IT application, which governs our interpersonal relationships.

The lower level is the neurologist's field of expertise (based on physicochemistry and electrical engineering), the middle level is that of the computer scientist, and the top layer is that of psychologists and psychiatrists and specialists in special applications. In our definition, other apps are, for example, driving a car, the higher levels of a language and the culture that goes with it, a science, perhaps philosophy? This is all in world 2′. The American philosopher Daniel Dennett (b. 1942) defines quite in this sense: *"The conscious human mind is something like a sequential virtual machine implemented—inefficiently—on the parallel hardware."* We define consciousness:

▶ **Definition Consciousness is based on the functions that are necessary for the pragmatic management of life in the world. These functions form our operating system of life.**
   A part of the functions is made conscious in a kind of app and assigned with language. This app is the consciousness in the narrower sense.

In terms of computer science, consciousness is nothing special. Some hard parts of the task have already been realized, for example in the self-driving car. What is special about consciousness for us personally is that we are "first-person shooters"—we are ourselves. The app of consciousness itself doesn't have to be localized in the brain or centralized in one area either. We have learned from computer networks that control and functions in a system can also be distributed. However, there is speculation that the brain region "*claustrum*" (lat. The front wall) is the area of integration of perceptions.

Consciousness is a well-defined informational system that is the focus of philosophical and scientific interest. The great actual philosophical and informational task in the science of mind, the "Big Challenge," is to understand a person's identity over a lifetime. Identity is the typical construct of a person out of himself and out of the environment and with a felt continuity throughout life, despite all changes and coincidences. Identity is what Plato had Socrates formulate in the *Symposium* ("the banquet"):

> *"Every single living thing, as long as it lives, is considered and called the same: e.g. a man is considered the same from his infancy to his old age. But although he bears the same name, he never remains the same in himself, but on the one hand he is always renewing himself . . . the one comes into being, the other passes away."*

The task today is to understand this account of Socrates in informational terms. How does the brand "I" preserve itself?

Computational Creativity and Directed Randomness   We have already mentioned the widespread opinion that a computer "cannot do anything new", *"because humans put everything into it"*. One experience that unfortunately seems to support this conviction is programming lessons at school. After all, the point there is to build little clockworks! But "real" software systems are not like that, not even simple programs. If we extend the above wisdom to the assertion: *"Computers can't do anything surprising, because humans have entered everything"*, nobody can really agree with that anymore.

However, we are thus conversely close to creativity. The computer scientist and computer art pioneer, philosopher and physicist Max Bense (1910–1990) wrote in 1965:

"Art is based on the frivolous nature of surprise".
Max Bense according to the "Spiegel", 18/1965.

We elevate surprise to the weak definition of creativity. Thus, the humorist in the quotation from Arthur Koestler also falls into the definition set, in the spirit of Arthur Koestler and Gilles Fauconnier. Even any sufficiently complex computation, though deterministic, can deliver "art" with surprise. With his first works of computer art, small random graphics, Max Bense horrified artists who felt that their creative possibilities were being competed with.

Max Bense goes further in our sense when he writes that a painter may know the subject of the picture he wants to paint, such as "Leda and the Swan" (Fig. 7.18), but it is only at the end of the painting process that all the "micro-aesthetic details" have emerged.

Thus, we have two levels of creativity:

(a) "Big" decisions, such as the subject of the painting in the first place, the degree of eroticism, the attitude of Leda, etc.,

Fig. 7.18 A work of art as an example by Max Bense of the "macro-aesthetic" and the "micro-aesthetic" traits. Image: Anonymous after Leonardo da Vinci. Around 1510–1515. Image: Leda and the Swan 1510–1515, Wikimedia commons, Web Gallery of Art

(b) "small" choices like Leda's hair, the details of the swan's feathers, the little flowers on the ground, the leaves on the tree, etc.

Both moves are more or less directed coincidence, the macro-effect more philosophically-contentually oriented, the micro-effect more unconsciously out of the noise. We model case a) in the tradition of Llull as double association ("bisociation") after the Hungarian-British writer Arthur Köstler, case b) loosely after the psychologist Dietrich Dörner and the American computer scientist Ray Kurzweil in the next chapter.

Central in both cases is chance - but it is not mentioned in all the discussions about creativity, and if it is, then rather negatively as uncreatively. "It" simply happens, it combines, it appears. Only when you want to map the process concretely in software, then you need chance explicitly.

The English language is partly to blame here: "randomness" is precisely not a creative and active word, whereas "Zufall", which strikes like lightning, is.

Due to their power and the enormous memory sizes, computers have an enormous potential—together with chance—to execute "creative" applications, such as

- Create" visual art, real and virtual, in two and three dimensions,
- Writing music,
- Composing poetry,
- Prove mathematics,
- Inventing engineering solutions,
- Telling stories and "understanding" stories, e.g. writing summaries,
- Writing and explaining jokes.

The last two points are particularly human and thus interesting but also problematic. Writing synopses and answering simple questions are undemanding tasks, but generating a story that is perceived as human is more difficult. By controlling by chance, however, the protagonists would have different names and roles in the second telling, and perhaps the story would even have a very different outcome.

Inventing jokes that are "good" in human terms is the supreme discipline of artificial intelligence:

For ambiguity, we have an approach with bisociation. But it also needs an understanding of irony, sarcasm, social boundaries and a comprehensive everyday knowledge. There are beginnings of this, but it is not yet possible for the German native speaker to fulfil the wish: to hear an Anglo-Saxon joke, simply not understand it although you know every word, and then be able to press the "help" button and have it explained! It's more possible to search the internet for jokes and show me the ones I should like based on my previous assessments of jokes.

Randomness and the computer come together directly and indirectly. Indirectly, the random character and stochasticity arise by themselves when dealing with large numbers of real objects—be it images, text volumes or customer data.

7.3.2   The Problem with Artificial Chance

"As a tool for picking something at random, there's nothing better than dice."
Sir Francis Galton, British naturalist and statistician, 1890.

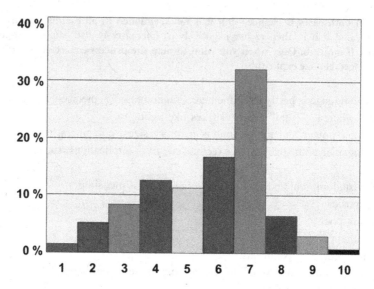

Fig. 7.19 Example of a distribution of human random numbers between 1 and 10. Image: own by report in imgur.com/gallery/wXSwdXr

Humans cannot simply produce random numbers. A simple example is asking a subject, *"Name a [random] number between 1 and 10!"* Fig. 7.19 shows a typical distribution of mentions. Boundaries 1 and 10 are disliked and rated as "non-random" in their boundary properties; in general, proximity to the boundaries seems off-putting. The most probable answer in the figure and according to various sources is the prime 7.

The distribution is far from the ideal equal distribution of 10% for each number.

If an injection of randomness is needed directly in a task, for example to compose music, one needs many random numbers, especially "good" random numbers, i.e. numbers that carry no or no unwanted correlations. Especially "good" random numbers require computational methods that perform a large number of similar random experiments to calculate something that would otherwise be impossible or difficult to determine, so-called Monte Carlo methods. Any deviation from ideal randomness would reduce the accuracy of the methods or produce spurious effects. The hardest demands on the quality of the randomness are made by cryptography. A key in a procedure must not be transparent, because it controls the whole secret conversation. The mechanism for generating it may be known, but it must be impossible to infer the inner state of the generator. There should be no pattern in a stream of random numbers that even tends to predict. This is also the requirement for digital roulette games, such as those of online casinos.

In our terminology, randomness is supposed to be completely directionless, neutral, clean, and without propensity.

All digits of the number space should be equally accessible.

In practice, there is also the demand for simple and economical production. We humans find it very difficult to tell from a numerical sequence of numbers whether it is "correctly" random or whether it contains trends. The computer finds this out quickly, but we also manage to do this with suitable visualization: If, for example, the numbers are plotted as a bitmap in a picture, '0' is white, '1' is black, we (i.e. our brain-computer, visual sense department, discovers) very easily see patterns. This is demonstrated by the pictures Figs. 7.23 and 7.24.

A curious counterexample, to which we do not look at the properties, is the number of the British mathematician David Champernowne (1912–2000), the Champernowne constant. In the decimal system it is the sequence C_{10}:

$$C_{10} = 0.12345678910111213141516171819202122232 4 \ldots$$

It is simply the sequence of all integers written one after the other! The amazing thing is that the frequencies of single digits, pairs of digits, triplets of digits, quadruples, etc. are identical to "real" random numbers—but it is obviously not a random number.

Figure 7.22 shows the typical dot pattern of a set of "very well random" random numbers without correlation. There are definitely areas which are darker, i.e. more densely populated, and those which are lighter. This corresponds to the well-known observation in one dimension, in rolling dice, that it is not impossible to roll "six" even three times in a row. There is a human problem with this. A random sequence with several sixes is no longer perceived as random. This is expressed by the cartoon in Fig. 7.20.

The joke of the cartoon has an entrepreneurial and a scientific side: The joke for the managers is the financial system of the Dilbert company, which has a random number generator to generate current or planned business figures (so-called Ballpark Figures). The scientific side demonstrates ambiguously that in the stream of random numbers, improbable things can appear—it's just random. If you want something to look like random for sure, you have to deliberately deviate from it and suppress something like "6 × 9". Figure 7.23 shows a set of points with humanized numbers, called "low-discrepancy random numbers".

Fig. 7.20 The finance department's random number generator. Cartoon of the series "Dilbert"from October 25, 2001. Andrews McMeel Universal

The human reaction and evaluation of total randomness is philosophically quite interesting, even compared to randomness in nature:

- Total randomness seems slightly unnatural ("6 × 9" is suspect),
- total order is boring (too much regularity means too little surprise),
- tamed chance is life (now and then even a six twice when rolling the dice makes it exciting, or a little trembling of the eyes at the doll PONG creates the impression of liveliness).

The extremes in nature are again something special, for example the total chaos (e.g. the fresh, chaotic lava field after a volcanic eruption) on the side of chance, and the fascination of crystals on the side of perfection or near-perfection.

The problem of getting good random numbers for computer applications is about as old as computers themselves. As mentioned, Alan Turing designed and enforced the first random number generator for the computer—he had already recognized the importance of randomness. After the first special devices, in 1951 a general computer, the Ferranti Mark I, received the hardware to generate twenty random bits from the noise of the electronics by program instruction. Generated with this randomness was the first computer-generated poem in history. Chance selected the sequence from the list of "romantic words", here the beginning:

JEWEL LOVE
MY LIKING HUNGERS FOR YOUR INFATUATION
YOU ARE MY EROTIC ARDOUR
MY FOND RAPTURE ETC

The originator of the love letter was MUC, the Manchester University Computer. Given the high value of computing time on the computer, it probably took a lot of humour to use the computer for a kind of poem—plus the foresight to perhaps go down in computer and literary history with it.

Since about 1940, the new Monte Carlo methods have been emerging, probably first conceived by Enrico Fermi around 1930. These are computational methods that are particularly suitable for the computers to come, and which require random numbers that have no interaction with each other whatsoever. In Popper's terminology, they must not have any "propensity", however hidden, or in the jargon of cryptographers (and magicians), the numbers must have "nothing up their sleeves".

The mathematical generation of pseudorandom numbers is already possible, but problematic:

"Anyone who considers arithmetic methods for producing [good] random digits is, of course, in a state of sin."
John von Neumann, Hungarian-American mathematician, 1951.

By 1950, people were thus aware of the importance of "good" random numbers, but generating them was problematic (Turing's hardware generator was error-prone).

For practical purposes one of the most famous and curious books in mathematics, indeed in general, was assembled: the book of "*1 Million Random Digits with 100,000 Normal Deviates*" (Fig. 7.21). The figure and the book are legally in a curious position as random digits. Wikimedia's licensing note for Fig. 7.21 reads:

> "*This work cannot be copyrighted. It is therefore in the public domain because it consists entirely of free information and has no original author.*"

After all, we had already forbidden the question of the exact cause when rolling the dice! The preliminary work for the book began around 1947 with a kind of "electric roulette", which provided raw numerical material (Figs. 7.22, 7.23 and 7.24).

The book is available in new edition today for about $200 and has become a cult favorite. Here are some of the 700 reader comments from Amazon.com:

- *To whom can I report the typos? The first '7' in the third line page 48 should be a '3'. The '7' there is not accidental. But otherwise, it's a good book."*
- *A wonderful reference book! But it's a pity that you don't have the numbers in order, then you would find the number you are looking for faster.*
- *A riveting book, suspenseful to the last page.*
- *Spoiler alert: The ending is 41,998. Read it anyway, you'll never figure out how it gets there!*

Fig. 7.21 A small section of the book "A Million Random Digits …". Image: Random Digits, Wikimedia Commons, RAND Corporation

73735	45963	78134	63873
02965	58303	90708	20025
98859	23851	27965	62394
33666	62570	64775	78428
81666	26440	20422	05720
15838	47174	76866	14330
89793	34378	08730	56522
78155	22466	81978	57323
16381	66207	11698	99314
75002	80827	53867	37797
99982	27601	62686	44711
84543	87442	50033	14021
77757	54043	46176	42391
80871	32792	87989	72248
30500	28220	12444	71840

Fig. 7.22 Pictorial
representation of a set of ideally
distributed coincidences. Image:
random 10,000, Wikimedia
Commons, Robert Dodier

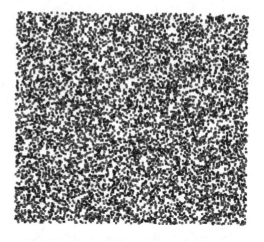

Fig. 7.23 Pictorial
representation of a set of random
numbers with low discrepancy.
Image: low discrepancy 10,000,
Wikimedia Commons, Maksim/
Dodier

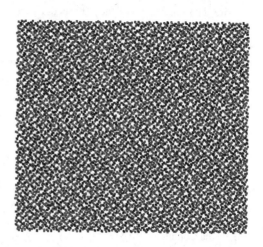

Fig. 7.24 Pictorial
representation of an internally
correlated set of random
numbers. Image adapted from
Bo Allen, boallen.com/random-
numbers. With friendly
permission

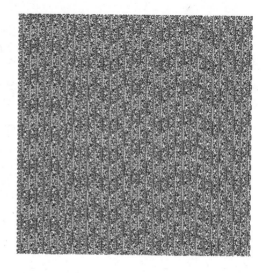

It is a pleasure to read the comments including from the young man who needed random numbers for a telephone survey with the book and found his future wife that way, a real "random woman".

Today there are many methods to generate random numbers via physics, classical ones like the dice since about 5000 years, exotic-curious ones (Fig. 7.25) as well as modern methods of highest quality. A problem of many mathematical methods is that the sequence of generated numbers repeats (too fast). The requirements are therefore highest certainty of non-repetition of the sequence of numbers supplied and highest possible uniform distribution or, in other words, maximum entropy (Fig. 7.26). Somewhat more delicately, the generation of random numbers can be described as the "generation of entropy".

The core of these processes responsible for physical randomness lies in the depth of atomic nuclei, in the thermal movements of a liquid or in the noise of moving electrons, measured as electrical voltage or as light pulses.

Curious was the professional application of the chaotic motion of liquid bubbles within lava lamps (Fig. 7.25) as an encoder for a random number generator. The computer company Silicon Graphics had built a generator with it: A camera produces an image of the lamp and from that a random number of 140 bytes for further processing. It is one of the nicest ways to generate random numbers while having a representative office object.

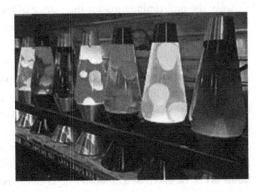

Fig. 7.25 Lava lamps. Lava lamps can be the randomizer for a number generator. Image: Lava lamps, Wikimedia Commons, Dean Hochman

Fig. 7.26 A modern random chip. Quantum chip from ID Quantique for security applications. Image: Company image ID Quantique, SK and Samsung (detail)

The modern random number generator of the Swiss company ID Quantique at the pencil tip in Fig. 7.26 basically uses the same physical principle as the random number generator of Alan Turing, namely the noise of electrons. In 1951, it was the noise of the anode current in an electron tube, today it is the noise in an LED whose generated photons are measured. It is the shot noise, which Walter Schottky discovered in 1918, and which we hear acoustically in analogy when raindrops tap on a tin roof. However, the technology has changed a lot since Schottky and Turing and has become many million times more sensitive: In the case of the chip, which is only a few millimeters in size, a few electrons and photons are sufficient, or even just a single photon, to record the coincidence.

Mathematics and computers today can produce excellent random numbers, but they are just deterministic. The mathematical construction of "safe" randomness has become a sophisticated branch of mathematics.

Mathematical software that creates randomness, like "real" physical randomness, is also philosophically interesting (Hehl 2016):

- The core of the software that generates randomness is deterministic-mathematical.
- The procedure and the program may be known or unknown.
- The procedure can specify boundary conditions for chance (a propensity).
- At least one element can prevent the output event from being predicted.
- Only the programmer and/or the starter of the program know this element.
- It is not possible to interfere with the course of the program. It would destroy the ideality.

We conclude that there is no difference between "real" randomness and "good" pseudo-randomness: they are events coming out of an impenetrable wall, behind which is the oracle of Alan Turing. The oracle is mathematically constructed or physically obtained; we have already mentioned a number of possibilities.

For low demands on artificial coincidence already the human being is sufficient as producer, but only for the first beginning. The movements of the mouse on the computer by humans can be taken as a random trigger (as a "seed" for an algorithm) to start software that generates a stream of pseudo-random numbers with mathematics.

The classic method of simply taking people "at random" from the phone book for a survey, for example, is not sufficient to get a good random set. Not in the past, when many people didn't have telephones, and not today, because many don't have a listing, perhaps even a landline telephone. Also, choosing all names starting with "H" does not give a random set!

7.3.3 Computer and Human Decide

"In every success story, you find someone who has made a courageous decision."
Peter Drucker, Austrian-American economist, 1909–2005.

The expression "to make a courageous decision" says it already: Some decisions are made in a state of some uncertainty. It is uncertainty due to the imprecision of the initial knowledge and it is uncertainty regarding the impact in the future. Nevertheless, decisions are made. About small things: *"Should I order these shoes?"* or about big things: *"Should I sell my business? Or invest in a different direction?"*

Three Scenarios of Decision with Chance: Group, Computer and Individual Human Being
We consider the decision-making process in three scenarios: Group decision (with leadership), decision by computer, and decision mental/rational of the individual.

To first understand the steps of the decision process, we map an external decision process, e.g. the management decision in a company by the "president" or CEO with his team. (Fig. 7.27). By distributing the decision process among the people in a group, the structure becomes more visible.

Here is a small fictitious example. The problem is to adapt the strategy of a company to the changed market situation. The current task is to determine the new strategy for the fictitious metal goods company MWAG:

Does MWAG want to build weapons again?

- *Consultant1: If we want to maintain sales, then we need a new line of business. I suggest new types of weapons. We already have a good patent on that.*
- *Consultant2: We should produce weapons again? That does not give a good reaction.*

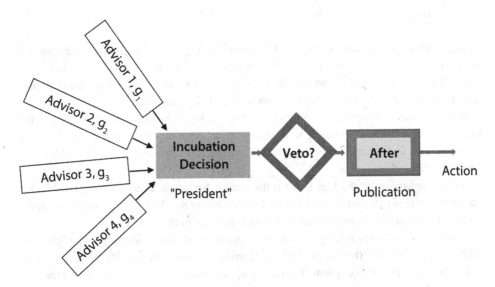

Fig. 7.27 The schematic representation of the decision-making process of a "president" with a group. In the "publishing" stage, the press office, the decision is subsequently justified. Image idea after Ray Kurzweil (2012) and Walter Hehl (2016)

- *CEO*: *I just read a report on that: Gun sales are up. We should be able to cut ourselves a slice of the cake.*
- *Major shareholder (veto)*: *I am against it. I propose the entry into ... etc.*
- *CEO*: *We do it this way. Our communication must highlight this in a positive way.*
- *Press release*: *MWAG will expand its activities. There will be no job cuts.*

The rationality of the process means dependence on information sources. External and internal knowledge bring the consultants, the importance of which is evaluated by the weight factors g_n. One of the sources is the knowledge of the president himself. The president does not need to participate in the deliberations of the advisors among themselves. At the incubation meeting the president decides. Possibly there is still the possibility of a veto—it may not go as easily as thought, the proposal is suppressed, the process goes back and another solution is sought. Finally, the decision is made. The execution of the action begins and in parallel the message is issued to the press, which provides a coherent argument.

The schematic suggests a precision that is often not present, because almost all fundamentals are inherently uncertain in direct information, and the possible choices amplify that uncertainty in effects. But uncertainty means room for chance!

This also applies to decision-making by computer. The computer is increasingly used to make decisions on a large scale, e.g. for

- the pre-analysis of applications for job advertisements in companies,
- a criminal's assessment of a possible leave of absence,
- the making of a cancer diagnosis.

All function blocks here correspond to software with information. Based on the limited information about the candidates and the uncertainty of the criteria, chance also has an effect. The rules for the decision are fixed, comprehensible and—at least for the programmer—visible. The feeling of the population is just the other way round: decisions by computer are seen as sinister, as the "power of algorithms". But human decisions are also problematic, arbitrary and volatile to boot! However, regulation by society (or at least the desire for it) is foreseeable, more important than general data protection: it is directly about decisions that influence fate.

The programmer (or the legislator) is the creator for whom everything is possible. He can prefer or reject, be just or unjust, have compassion or not. Due to the uncertainty in the input data, the decision contains chance, which evolves in the future.

The human decision-making process is even less determined: Determinacy ("decision with certainty") is the (boring) exception. Human decisions are usually "decisions with risk". The decider does not know the true values for sure, is not sure of himself. Here is a bon mot about it:

"People like to say 'what exactly do you mean by that?' I would answer 'I mean it, but not exactly'."
  Jean-Luc Godard, French-Swiss film director, born 1930.

The key word is fuzziness or vagueness. The word means according to Wiktionary

  "Fuzzy: covered with fuzz, or a large amount of tiny fibers".

  Effectively, fuzziness is a transfer of "noise" from physics and electrical engineering to information:

Fuzzy information is inaccurate, inappropriate, false or even deliberately falsified information along with the "true" information. In this sense, the whole Internet is unfortunately "fuzzy".

  For illustration purposes, various blocks are drawn blurred in Fig. 7.28 and traced with the "paintbrush" for this purpose: There is definitely correct and exact information, but much is blurred. This concerns especially unconscious processes and unconscious "advisors". In contrast to the personified functions of Fig. 7.27, the functions of this block diagram are thought of as functions in the brain, but without implying spatial localization. The left side of the figure is primarily unconscious; it is knowledge mixed with emotion. Only the gaining processes come to consciousness.

  Electrical-physical noise and informational noise, the whole variety of thought patterns and pieces of information, meet in the brain. Spontaneous electrical noise is a feature of our

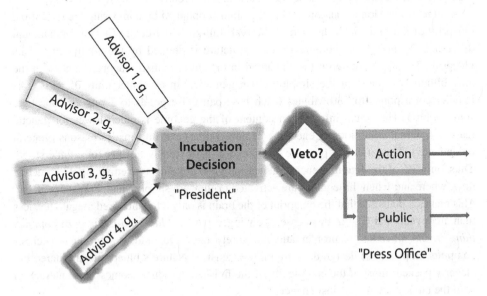

Fig. 7.28 The schematic representation of individual decision-making with an emphasis on uncertainty or "fuzziness". The blocks are meant here figuratively as internal, unconscious psychological functions

brain, whether it is busy or not (Nicolic 2018). They are largely small spikes, but occasionally larger, avalanche-like events. Inject a narcotic and the noise barely changes. It is fundamentally part of the living brain, from the molecular to the psychological level.

In the spirit of our theme of "randomness", an essential property of brain activity is the uniqueness of each action. An elementary computer action always produces the same result with the same input, differently in the brain (Segal 2019):

> **"In the brain, even if you choose the exact same stimulus, the response varies from trial to trial."**
> **Michael Segal, Israeli computer scientist, born 1972.**

Incidentally, in a computer system with a large number of interacting processes, or even when working at the limits of computational accuracy, it is no longer certain that the results will be identical.

At the IT level of the brain (Fig. 7.17), we find concepts similar to those in modern artificial intelligence, such as the use of pattern recognition and deep neural networks. In addition, there is the characteristic software for the inner view, the consciousness, because we are inside the computer. But the basis of biological neural networks in the brain, with its incredible parallelism and inherent stochasticity (the "funkiness"), is different and much is still not understood. The extreme case of "sparkiness" is quantum fluctuations. The influence of quantum effects is not excluded; but in the normal case, millions and billions of electrons are involved in the firing of neurons. The architecture of the brain is that of a high-performance computer in a curious, slow and inherently unreliable technology.

Another dimension of randomness in the brain is obtained by comparing the size of the blueprint of the brain with the size of the real, full-grown structure. It is a fundamental difference. Naturally, the complexity of a structure is defined by the complexity of its blueprint. The brain's blueprint is contained in the entire human blueprint, that is, in the 3.27 billion base pairs of the simple human genome.[3] In computer units of mass, this initially corresponds to 780 Mbytes (each base pair corresponds to 2 bits and one byte stores 8 bits). The actual information content of the genome is much lower for genetic reasons (internal conditions of encoding or non-coding) and informatics reasons (general redundancy). This effective possible compression is obviously not yet fully understood. Thus, only 3% of the genomes encode genes and yield proteins, the remaining 97% encode how, where and when these genes are activated or they are simply "junk" (Watters 2019). This makes it plausible that the blueprint of the brain is only a few hundred megabytes, less than a conventional CD: in this sense, the winged phrase *"the brain is the most complex thing in the universe"* is problematic, for surely there are industrial products such as computer chips whose design data *are* more extensive! Nature's blueprints are incredibly cleverly packed; most of the complexity of the finished "product" comes from interaction with the environment or is just chance.

---

[3] Each chromosome is counted only once, not twice as in a body cell.

The situation is different if you relate the complexity to the finished product "adult brain": Here you have almost 100 billion neurons, each interconnected with up to 10,000 other neurons via a network (or several superimposed networks) with the 10^{14} to 10^{15} synapses as connections. In the growth phase and during life, an enormous amount of information is created by the building action of the DNA by means of biochemistry on the physical side and by lifelong learning on the mental side. A rough calculation gives many TeraBytes or even PetaBytes for the finished construction, however with unknown redundancy. In the construction of the brain, nature shows a mastery of compactness. A large part of the total information is random, but in different forms:

- "fossil" coincidence in individual DNA from evolution, almost becoming a law of nature.
- Individual randomness of life, shaped by the partly random growing, living and learning.
- Randomness of lifelong interaction with the environment, regular "randomness".

Here the term *randomness* is appropriate as a designation of many random little things, but still with great possible effect at one or the other randomness. Our brain contains a lot of randomness in the given static structure and it works dynamically with sparkling randomness.

Philosophical Borderline Case of Decision: The Donkey of Buridanus
The extreme case of the decision task is well known in philosophy as the problem of Buridanus. It belongs to the vocabulary of the educated as this excerpt of a dialogue from the American TV series *The Big Bang Theory* shows:

- **Amy:** "What are you doing?"
- **Sheldon:** "I'm just contemplating Buridan's donkey."
- **Amy:** "I understand. Then I'll leave you be."
- **Sheldon:** "What, you are familiar with the reference?"
- **Amy:** "But of course."

It's about the problem of deciding when, rationally, all arguments are equally valid. This idea is found in the story in a variety of fictional situations:

In the common version, a donkey starves to death between two equally appetizing haystacks (Fig. 7.29). Jean Buridan (1300–1358) describes the problem several times: a dog starves to death with him because it cannot decide between two sources of food, a hiker despairs at a fork in the road, and a sailor in distress at sea who will not throw his cargo into the water. In Aristotle and in Dante a man is supposed to choose between equally attractive drinking and eating, in the Persian philosopher Al-Ghazali between two equally juicy dates (or two glasses of water). In addition, there is the cosmic-physical variant in the Greek philosopher Anaximander (610–546 BC). He imagines the earth suspended in space equidistant from everything else in the universe. Now all directions are equal and the earth therefore remains free-floating and at rest.

That the earth is in the center of space, because no direction is distinguished, is a good physical argument, if one allows only one earth. This is thus a logical inversion of the Copernican principle.

Sheldon, the hero of the TV series, even knows that the idea of the fable goes back to Aristotle. Aristotle makes fun of the dilemma; it is as ridiculous as

"a man as hungry as he was thirsty, and being between eating and drinking, could not stir and come to death."

The Dutch philosopher Baruch of Spinoza (1632–1677) also makes fun of those starving at a laid table; he probably deliberately linked the problem of Buridanus to a donkey. That it is a starving donkey rather than a starving man suggests a mockery. Jean Buridan, French philosopher (1301–1359/62), wrote in 1340:

"If two actions are judged to be the same, then the will cannot overcome the dead point. All it can do is postpone the decision until the boundary conditions change and the right action is clear."

The opinions of historical philosophers diverge, even confuse: for one, free will solves the problem, for another it just blocks it. The Persian philosopher Al-Ghazali (1058–1111) can be interpreted in a modern way when he writes around 1100:

"Suppose there are two similar dates before a man. He longs for them, but cannot take both. But he will surely take one, because he has a quality in him that can also distinguish between two very similar things."

In technology, for example in the computer, such self-blocking situations do occur. *The "property in us"*, which can distinguish between very similar things and which can resolve such "deadlocks", is chance. For example, one can assign random numbers to the *"dates"* and the higher number wins (is eaten).

In the mind, no artificial assignment of randomness is necessary. There is always randomness in the form of noise, as shown above. A noise spike and a random pattern decide. According to Fig. 7.27, if no advisor clearly dominates, then a random argument or a noise spike wins. The noise of the brain is the property in us that Al-Ghazali paraphrased; the small randomness unblocks.

Physical Limit Case of Decision: The Dome of NORTON

"Newtonian physics is deterministic, sorry Norton".
Gareth "Gruff" Davies, physicist and inventor, 2017.

Fig. 7.29 The donkey of Buridanus. Here England before the Paris vs. Berlin election. Image: Caricature by Alfred Lepetit, 1870. (Source: Ane de Buridanus La Charge, Wikimedia Commons, Parismuséescollections)

Fig. 7.30 A sphere rests on the top of the Norton dome. A thought experiment in decision-making. Image: Norton's dome, Wikimedia Commons, Witherk

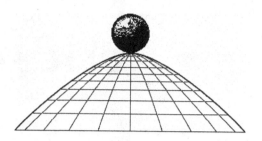

In a 2003 paper, chemist and philosopher John Norton suspected he had uncovered a kind of quantum mechanical indeterminism in classical Newtonian mechanics, a kind of active randomness. Norton was wrong, but he became world famous with his imagined object.

Whether the teaching of Aristotle, Buridanus, or Newton, it always takes a force to come out of rest. The difference between Aristotle, Buridanus and Newton is motion. Motion ceases by itself according to Aristotle and Buridanus, but according to Newton it continues without cause. A body remains at rest according to all three if no force acts on it. Galileo Galilei, by the way, still has one foot in antiquity; he still thinks circular motions also simply go on "by themselves and forever" like the planets on their orbits.

If you let go of a (smooth) ball on a (smooth) slope, it starts moving downwards at an accelerated rate. Let us now consider a sphere on an egg-shaped body and let us move it higher and higher on the egg until it reaches the top (let the egg be perfectly symmetrical for this purpose). We get approximately the picture of Fig. 7.30. What about the egg? The philosopher Norton devised a special shape, but it does not matter for the problem, as it turned out:

The infinitesimal environment of the highest point is decisive.

- If the ball were balancing on the highest point of a cone, it would roll downward with constant acceleration at the slightest vibration. But the infinitesimal environment is a small piece of plane.
- If the sphere were on top of a sphere, it would be on a small piece of tangent plane—but just a little away from the center, and there would be a preference, and thus a defined downward.

Norton came up with a special curve that forms the outline of the dome: This curve is particularly flat at the top (the radius of curvature of the curve is infinite at the top).[4] Practically, the upper sphere is even positioned a bit more stable on the tip of the dome than on a normal large sphere—contrary to Norton's assumption that it can run loose by itself. The rest of the paper is physico-philosophical pseudo-discussion; in modern parlance, the Norton dome is a "red herring" (Davies 2017):

---

[4] The first and second derivatives of the contour function are zero at the top, only the third derivative exists.

"A dead red herring was often used to throw a hunting dog off the scent and test its nose."
(*Urban Dictionary, pulled 2020.*)

Let us rather go back to the eleventh century to the pragmatic Persian philosopher Al-Ghazali. We humans have the sense for realistic results and that means here for the influence of coincidence. In the case of the sphere, for one thing, the roughness of the surfaces involved is random in detail. This roughness works weakly against the tilting of the small sphere. On the other hand, it is the random fluctuations of the two spheres, against each other and with the base. The randomness will determine the direction of the slide. The highest point of the Norton ice is an artificial singularity of the ideal configuration, which chance pragmatically resolves. It is the difference between real physics and ideal mathematics.

7.3.4   "Free Will" and Chance

"Free will is the brain's ability to take action and make independent decisions based on internal motivations, not reactive reflexes (escape reflex) or external compulsion."
Peter Ulmschneider, German astronomer, born 1938.

The "Free Will": The Technical Functioning

The above quote is probably a widely accepted definition, but the crux of the pudding is "the inner motives", and both the attribute "inner" and "motives".

We remain at the level of a simple block diagram. Figure 7.27 is just the metaphorical picture of the inner motivations from an IT and process point of view. If one considers the brain as an information-processing system without supernatural parts, then one must draw a logical circuit diagram to understand it.

Since we are dealing with physical IT operations and not (solely) with philosophy, the picture is also a temporal sequence of steps. This sequence was measured in the Libet experiment by the American physiologist Benjamin Libet (1916–2007). According to Wikipedia "Libet experiment", in a simple experiment (the subject has to press a lever at some point, according to his "free" choice), one measures when the brain reports the action through its signals, when the subject clearly feels his decision, and when the hand really moves. The latter mentioned point in time is the reference point zero:

- −1050 msec, if the subject planned the action ahead,
- −550 msec, if the action should be executed spontaneously,
- −200 msec until (in both upper cases) the decision became conscious.

In addition, there was the possibility of a veto in the 200 msec before the decision, here consciously, but also in principle unconsciously possible.

The Libet experiment says nothing directly about philosophical "free will," but it does confirm the validity of a principle picture like the one in Fig. 7.27.

However, at this level of simplification, there are few other natural options!

According to this, "free will" is, in informational terms, a confluence of rules and chance, and of psychological and factual experience and chance. In the picture, chance is indicated by the fuzzy contours; factual rules and information are more sharply drawn, psychological and emotional influences are "fuzzier". But even hard numerical values can be misinterpreted—e.g. a random error in the decimal place for larger amounts as already experienced by oneself with old price indications in pesetas or lire . . .

The American biologist Anthony Cashmore, a member of the Akademie der US Nationalen Akademie der Wissenschaften, sees a special, genetically based mechanism for our (illusory) free will as an evolutionary function (Zyga 2010): Actually, free will does mean a kind of foresighted organ that, consciously or unconsciously, looks into the future and compares virtual alternatives and then decides. What we think of as free will is just the unconscious reflection of the subconscious organ. This brings to mind the pioneer of the unconscious, the Austrian psychologist Sigmund Freud (1856–1939). Anthony Cashmore writes:

"Freud was right to an extent [regarding the importance of the subconscious] much greater than he imagined."

Let's put it this way:

Free will as a conscious feeling thus relates to this unconscious mechanism just as the whole of consciousness relates to the unconscious software of the operating system of our practical life.

To process impressions that lead to decisions, we humans have subroutines that work inside us and help make (mostly) helpful decisions based on experience. Psychology calls these aids heuristics. We have already learned about metaheuristics. Metaheuristics are general solution procedures with uncertainty. Heuristics are procedures for arriving at specific, workable solutions with limited knowledge. Uncertainty acts like chance (or even means chance). If in a heuristic the uncertainty goes to zero, we get an algorithm again. On the other hand, at the other end of the spectrum, we have defined the ability to obtain good solutions even in the presence of uncertainty as intelligence.

But there are many systematic sources of error for these inner decisions. Psychology calls them *cognitive distortions* in perceiving and judging and keeps a constantly growing list. Table 7.1 lists a few examples of "cognitive biases" from Wikipedia.

The English Wikipedia lists a total of about 190 (!) biases. Many psychologists and sociologists have been able to perpetuate themselves in their disciplines with their "own"

Table 7.1 Some cognitive biases that can influence decisions. Excerpt from the corresponding lists of the German and the English Wikipedia article

Agent detection	The presumption of the presence of another, unseen being. An evolutionary relic.
Apophenia	Recognition of apparent patterns in random
Confirmation error	The tendency to select information to meet one's expectations.
Cryptomnesia/False memories	Fantasy as memory or, conversely, memory as fantasy.
Halo effect	The unwarranted inference of known properties to unknown ones.
Hot hand effect	The random accumulation of success in gambling is perceived as a "lucky streak".
Illusory correlation	The erroneous perception of a correlation of two events.
Contrast effect	The more intense perception of a piece of information when it is presented with contrast information.
Control illusion	The false assumption that you can control random events through your own behavior.
Pareidolia	Recognition of apparent objects in pictorial structures. A special case of apophenia.
Sources interchange	Confusion of memories.
Backshopping error	The distorted memory of one's own predictions after the event has occurred.
Semmelweis effect	The reflexive rejection of novelty.
Availability heuristic	Whichever is associated first wins.
Von Restorff effect	Something special or isolated makes a stronger impression.
Truth effect	The tendency to attribute greater truth to statements that have been heard before than to those heard for the first time.
Probability ignoring	Incorrect assessment in situations of uncertainty.
Zeigarnik effect	Incomplete and interrupted tasks are better remembered than completed ones.

bias. The Swiss author Rolf Dobelli (born 1966) has written quite successful books about it (Dobelli 2011).

Many of the effects in the lists are well known. We want to comment on three effects because of their philosophical importance and because of the connection with coincidence, the "agent detection" and the "apophenia/pareidolia" and—not from the Wikipedia list— the "synchronicity" of the Zurich psychologist CG Jung.

Agent Detection, also known as *Hyperactive Agency Detection Device (HADD)* or "exaggerated suspicion of an acting being" is a psychological basis of the "watchmaker paradox". It is the tendency of animals, including us humans, to quickly see the activity of an (intelligent) being behind a trace or shadow, even if the clues are vague and it could also be coincidence.

The evolutionary reason for this is the dangerous situation in nature. The imprint of a lion's paw should not be interpreted as a harmless random hollow in the sand, and the cracking of a branch in a thicket should not be interpreted as a random sound. It is not harmful to be alert once too often, but very dangerous once too little. Great forms of "agent detection" are, of course, the acting gods in religions, elaborated into more or less detailed figures. In this sense, this effect is a relic of evolution and forms the evolutionary psychological basis of belief in gods—which of course could exist anyway (Hehl 2019).

Apophenia is the false tendency to see apparent patterns and relationships in random things and to attach special significance to them. It is the source of much superstitious wisdom. *Pareidolia* is the special case of this, the fallacious identification of familiar figures in things and patterns. The term apophenia originated in the clinical picture of schizophrenia, but is, if within limits, a normal process and source of creativity during incubation. This is especially true when the relationship found makes sense, as in the discovery of Auguste Kekulé's benzene formula, or even the theory of continental drift by Alfred Wegener around 1915. Wegener was struck by how well the coastlines of Latin America and of Africa would fit together if they were shifted and rotated a bit. Another accidental and thus wrong scientific example is shown in Fig. 7.31.

It is the fictitious face in this image taken by the *Viking1* probe in 1976 of the Martian plain Cydonia from an altitude of 1800 km. After its release, it became famous as the "Martian Face", and the hills next to it as the "Inca City". Rumors arose that it was the work of extraterrestrial visitors, comparable to the great Sphinx of Giza. Thirty-one years later, the *Mars Reconnaissance Orbiter* shows the mountain in high resolution. It is an ordinary mountain with a random shape. The interpretation of the first photo was a pareidolia. Too late! The Martian face is already part of pop culture.

Zurich psychologist Peter Brugger writes in 2012:

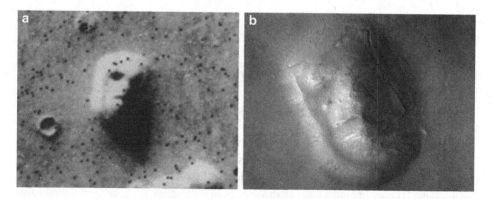

Fig. 7.31 (**a**) The fictitious Martian face as an example of pareidolia and apophenia. Image: Martian Face Viking cropped, Wikimedia Commons, NASA Viking1. (**b**) The same mountain in better resolution and different lighting. Image: HiRise Face, Wikimedia Commons, NASA/JPL

"We can't even perceive chance per se. We can only perceive what stings the eye, what stands out from the noise in the background - chance."

We notice only what penetrates the "covert inhibition" mentioned above, and that is what we interpret. In a paranoid psychosis, this quickly becomes horrifying.

Another fictitious perception in coincidence has entered cultural history through the Swiss psychologist Carl Gustav Jung (1875–1961): the effect of "synchronicity". By this CG Jung understood two events, an inner one (like a dream) and an outer one (like a misfortune), which he saw connected by a special relation, just "synchronistic". The inner event is the dream of a person, the outer one a misfortune that happens to this person or to a close person—if possible simultaneously. In no case should the dream occur after the misfortune. Experiencing such synchronicity can certainly be very impressive and compelling, but if there is no causal relationship, it is coincidence. CG Jung held a real force—beyond physics—responsible for such relationships. This part of his work is pseudoscience. Physicist and Nobel laureate Wolfgang Pauli (1900–1958) actively participated in the pseudoscientific discussion in a famous correspondence with CG Jung (Meier 1992). Synchronicity is neither science nor did it make it to the mentioned lists of cognitive distortions in Wikipedia as a psychological effect and deceptive feeling.

We have thus analysed—as defined above—the brain's ability to decide something on the basis of inner motives:

It, i.e. our mind or "the Mind", is technically an IT system that functions fuzzily with randomness and with countless distorting programs. Thus, the mind is the object of research of neurology, psychology and computer science.

The "Free Will": The Inner View: We as First-Person Shooters

"Man is a masterpiece of creation also for the very reason that he believes, despite all determinism, that he acts as a free being."
Georg Christoph Lichtenberg, German physicist, 1742-1799.

Georg Christoph Lichtenberg is a brilliant physicist and biting aphorist of the eighteenth century. The following quote is from the German philosopher Arthur Schopenhauer (1788–1860). It's a bit long and originally in classic philosophical German, but it's brilliant. The most important parts are highlighted in bold, and the italicized parts are emphasized by Schopenhauer himself:

"Hence the question remains: is the will itself free?—Here, then, the concept of freedom, which had hitherto been thought of only in relation to *ability*, had been placed in relation to *willing*, and the problem arose whether willing itself was *free*. But to enter into this connection with *willing*, the original, purely empirical, and therefore popular concept of freedom shows itself incapable, on closer examination. For according to this, "*free*" means "*according to one's own will*": if one now asks whether the will itself is free, one asks whether the will is according to itself: which is self-evident, but by which nothing is said. According to the empirical concept of freedom, it is said, "I am free when I can *do what I will*": and by "what I will" freedom is already decided. But now that we ask about the freedom of *willing* itself, this question would accordingly be put thus: **"can you also *will* what you will?"**—which comes out as if the willing still depended on another willing that lay behind it.

> **And if this question were answered in the affirmative, the second question would immediately arise, "Can you want what you want?" and so it would be postponed ad infinitum, in that we would always think of an intention as dependent on an earlier or lower one, and would strive in vain to attain in the end by this means one which we would have to think of and accept as dependent on nothing at all.**
> **Arthur Schopenhauer, German philosopher, 1841.**

Schopenhauer thus proved quasi-mathematically that the question of free will is meaningless. It is like opening a matryoshka doll, which again contains a doll, which again contains one, and so on. (Fig. 7.32). The point of decision is pushed further and further. His work won a prize from the Norwegian Society of Science, but the main message was probably more of a bon mot for the public. Einstein made it famous:

> **"I do not believe in the freedom of the will. Schopenhauer's word: "Man may well do what he wants, but he cannot want what he wants", accompanies me in all situations of life".**
> **Albert Einstein, Swiss-German physicist, 1936.**

Fig. 7.32 Matryoshka dolls illustration of Schopenhauer's proof of unfree will. Image: Matryoshka dolls, Wikimedia Commons, Stephen Edmonds

Emotionally, free will is a process that comes from the inner nothingness, just like that, without causality. But we deny that the action would be coincidence, we mostly justify it (pseudo) causally, sometimes it is an "I want it" without justification. An example is that smoker who says:

"Of course I could quit smoking anytime I *want, but I don't want to!*"

If a decision is made causally based on constraints or by accepting chance, neither is "genuine" free will. "Causal" is an algorithm and chance is an unknown and uncontrollable force. We are trying to define (Hehl 2019):

▶ **Definition   A system S has free will if there is in it a subsystem S_2, called *free will*, which, detached from the rest of the system S, judges all influences and makes an otherwise free decision.**

System S_2 is affected, for example, by the smoker's thought *"now a cigarette would be wonderful"* but also *"you've seen the model of the tarred lung."* Actually, system S_2 is now in the role of the free subject that should decide. We need to go further in the definition with this:

▶ **Definition (continued)   System S_2 has free will if there is a system S_3 that decides freely in S_2; S_3 is free if it contains a free system S_4, and so on.**

It is exactly the iterative definition Schopenhauer points out! If one continues the process seriously in the material brain, the possible ranges would become smaller and smaller down to atoms (and chance). Otherwise, the iteration goes on and on. There is no free will in the strict logical sense. It only exists in a felt sense through the app "consciousness", but that is not free.

For the system S_2 (and the following ones) there is an illustration and a classical name, which was introduced into the philosophy of mind especially by the American philosopher Daniel Dennett (b. 1942): the homunculus, Latin for "little man". In the late Middle Ages, the homunculus was an artificially created little human being; the physician and alchemist Paracelsus von Hohenheim gives concrete instructions for its creation in 1538.

Dennett wanted to take the classical dualism matter-mind in the style of Descartes *ad absurdum* and show that a dividing line between matter and mind does not work. He called it "Cartesian theatre" (Fig. 7.33).

Although, of course, no one believes in the real existence of the human being in the brain, but it is the implicit solution of the body-mind problem: the eye casts an image of the outside world on the retina. The homunculus looks at it and interprets it. But in the process the task repeats itself: the homunculus again needs a homunculus in its brain in order to interpret the image that is external to it, and so on, in indissoluble regression.

But a hidden human is a psychological concept that we use again and again when we anthropomorphize things and talk to their homunculus. When the car won't start "*If you let*

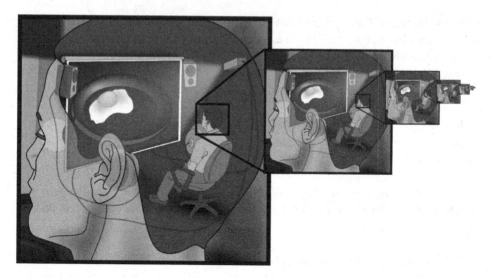

Fig. 7.33 The Cartesian theatre of Daniel Dennett, infinitely repeated. Image: Infinite Regress of Homunculus, Wikimedia Commons, Original Jennifer Garcia/(Reverie)/Pbroks13/Was a bee

me down now, then . . .", when the software quits *"Now come on back . . ."*, but also when a programmer describes the way his program works *"Now he's getting the data, now he's checking it, now he's passing it on to Alice, there's the bug!"* It is homunculus language, in the last case even justified, because "he" is really doing something and is almost human in doing it, the computer!

Emotionally, we are "ego-shooters" in life (from the Greek/Latin *ego* 'I' and English *'shooter'*). In German, the term is a bogus anglicism like "Handy" and "Beamer" and refers to video games in which the player acts from a first-person perspective in the three-dimensional world. In games with Colts, machine guns, and cannons, the view of the world is, of course, over the barrel of a gun. (In English, this is called first-person shooter games). The two first-person shooter images 7.34 illustrate this.

Image: Ernst Mach interior perspective, Wikimedia Commons, image-PD-old. (**b**) Action scene in the first-person shooter video game "Red Orchestra". Image: Red Orchestra25Shot0239, Wikimedia Commons, NASA/JPL. CC Attribution 3.0 Unported.

They are also contemporary documents: The drawing "Selbstanschauung" of the nineteenth century physicist on his couch (Fig. 7.34a) and the scene from the video about the rear sight seen in the twenty-first century (Fig. 7.34b). The two images symbolize the great problem of dualism. In dualism we have the brain here and the ego there. The pictures show what the *ego* sees, and the *ego* is the computer: this is how a computer sees the world.

For the sake of originality, let us mention a middle ground between the technical and the philosophical view, already noted by Daniel Dennett: the stages of homunculi interpreted as computer technology. Large software systems consist, as already mentioned, of many layers and, in addition, of many subprograms and sub-subprograms that call each other.

Fig. 7.34 (a) The first-person perspective. Drawing by the physicist Ernst Mach from 1886

"Higher" programs execute "higher" functions, calling lower programs with lower, simpler functions down to the elementary operations—through the neurons in the brain, or the transistors in the digital computer- and the introduced randomness. If the homunculi are interpreted as beings that are less and less intelligent downwards and simpler and simpler, the (now only finite) regression of the homunculi becomes the diagram of the software of our brain.

The German philosopher Immanuel Kant discusses the problem of free will (and morality) at length. Thus, he writes (Wagner 2005)

> **"I now say: every being that cannot act otherwise than under the idea of freedom is for that very reason truly free in practical respect, i.e. all the laws that are inseparably connected with freedom apply to it, just as if its will were also valid in itself and in theoretical philosophy, declared to be free."**
> **Immanuel Kant, Grundlegung zur Metaphysik der Sitten, 1785.**

This means that Kant understands that we cannot help but act as if we are free. And that we do not even understand why we act as we do, he says here (Kant 2019):

> **"The actual morality of actions therefore remains entirely hidden from us. Our attributions can only be related to the empirical character. But how much of this is to be ascribed to the pure effect of freedom, how much to mere nature and the unculpable defect of temperament, or to its fortunate constitution (*merito fortunae*), no one can fathom."**

The enumeration matches the influences on a decision in our diagrams Figs. 7.27 and 7.28; it even includes chance as *merito fortunae*, as luck.

In classical theoretical philosophy there is no explanation of how the "spiritual" comes into being. Matter and mind are separate worlds, and we feel it so! The explanation of the emergence of intellect, soul and consciousness as emergence from information technology was and is a tedious way against felt freedom. We have had to learn it over the last 70 years and are still learning. Today we know we are not free, but with a little luck we think we are. Chance and the noise of our mental machinery are also obscured and disappeared somewhere in our sense of freedom. But this also makes us responsible and brings us to the third view, the external view of the individual.

The "Free Will": The External System View

> "Thus, the first intention of existentialism is to put every man in possession of himself and to impose upon him the total responsibility for his existence. And when we say that man is responsible for himself, we do not mean to say that he is responsible for his strict individuality, but for all men."
> Jean-Paul Sartre, French philosopher, "L'existentialisme est un humanisme," 1945.

For Jean-Paul Sartre (1905–1980), human life consists primarily of chance, and it begins with a great chance: birth. We have seen that inner chance plays a role in our decisions, on the one hand in the form of the uncertain "noisy" information, but also directly as random combinations. To do this, we think of a boundary drawn around an individual that corresponds to the surface of the body. We can tell exactly where our hand is that belongs to us. We have included chance within us.

In Fig. 7.35, the three diamonds symbolize three people; the diamond symbol, a diamond in English, is the symbol for a decision in flowcharts. The arrows in the diagram

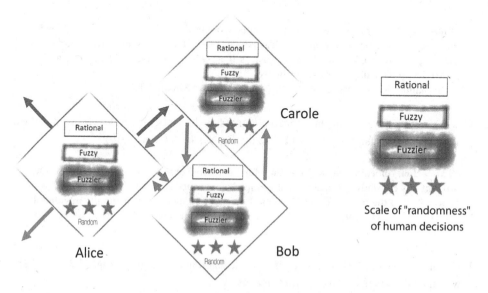

Fig. 7.35 Human relationships. People schematically as sources of decisions with more or less chance. The inherent coincidence belongs to the human being

illustrate the relationships between the people and to the outside. The inside of the diamonds symbolize our insights about decision-making. In particular, the stars inside remind us of the influence of invisible chance within us; in addition, there is visible chance in the outside world. The totality of influences, lawfully necessary as well as coincidental, is what we call fate.

The definition of free will seen from the outside is thus simple:

▶ **Definition   Every decision that is made by a rhombus without external coercion, we consider as freely made and the rhombus is responsible for it. Exceptions are decisions that are visibly caused by chance from the outside.**

The definition is independent of the question whether free will definitely exists or not. The Belgian physicist David Ruelle (b. 1935) reports that Werner Heisenberg remarks that the idea of free will of other people is not a problem at all (Ruelle 1994). Indeed, both are conceivable for us, namely that the other just acted "freely" or had to act that way. With the notion of the homunculus we understand this: To the outside world, it is indifferent whether the "Descartes ego", the "homunculus ego" or the modern psychological ego have decided.

So we are officially partly responsible for the inner noise—the little coincidences inside us and the resulting actions. The noise in the brain tips the scales for some decisions we make, and we can't help it. But we, as a system, do have a responsibility. This illusion of responsibility also has evolutionary advantages, at the latest in the social sphere, when we consider the effect of an action.

We also have no problem in this framework to consider the other as free as we feel free: What wine we choose to eat may be random, but we are responsible for it, on the other hand, not if the wine has "cork" and we lack the sense of smell to detect the smell of cork. The responsibility ends when the coincidence becomes visible.

Again, when we talk about propensity and "inclined" chance, we are responsible for the invisible inclinations as well. We are this sum together with our genes, our chemistry and our life course so far with everything we have experienced and learned, like "white wine to fish, red wine to roast." We can also use the term propensity to define character, namely as the sum of all our propensities. That is what we are, and for that we must answer.

Based on this, one can now develop the usual theories for state and coexistence.

The "Free Will": The Overall View

> "There's no such thing as free will, but it's better we believe in it."
> Title of an article in The Atlantic, June 2016.

> "Is free will an illusion? Don't trust your instincts when it comes to free will or consciousness, experimental philosophers say."
> Title of a Scientific American article, November 2011.

The ideas of free will still get mixed up today. The main reasons are the black and white thinking in determinism versus indeterminism and the reluctance to give up perceived freedom. We have two main ideas introduced into the discussion:

- The role of system boundaries.
- The importance of chance.

Thus, it is clear that freedom of will in the naive sense ("*I feel it*") or naive philosophical sense ("*there is spirit in us, not matter*") is illusory. It is unscientific today to believe in freedom of the will. And still the statements are valid:

- It is neither mere determinism nor indeterminism.
- It is not predetermined what will happen.
- It is not pointless to hold man responsible for his actions.
- There is nothing supernatural in decision-making (and in all consciousness).

The fuzziness of the large amounts of internal information (and algorithms) with inherent randomness, but also explicit randomness in the noise of the neurons, ensure that development is not determined. Even identical twins develop differently. Using Alan Turing's concept of the oracle, the source of randomness, we can say quite figuratively: We are also an oracle.

Chance creates alternatives and defuses or prevents determinism. If the world starts a second time with the same conditions, then a different course of the world results (Earman 1986). But chance is certainly not subject to our control: chance does not mean free will.

The sketches in the section, such as Figs. 7.27, 7.28, 7.33, and 7.35, will help to discuss the ideas. Seeing the system boundaries is essential. The author has learned how tedious it is to understand pure abstract texts; hence the recommendation:

> **"Never trust a writer who wants a complex philosophical problem to be understood without a chart. He's like a programmer on a team who just programs away without caring about the others. And in the process it is likely to overlook something himself."**

7.3.5   Summary of the Chapter

Creativity is historically a divine or semi-divine activity; as a field of psychological research, creativity has only existed since 1950. Creativity is closely linked to chance, visible or invisible. It is the way "the new comes into the world" (Klaus Mainzer 2007). Chance is most visible and most gratifying, operating in serendipity, in the serendipitous, unexpected find. Creation can occur unexpectedly or be attempted according to plan. We analyze the process of creation of an idea or invention. In doing so, we follow the early analyses of the physicist Helmut Helmholtz and the mathematician Henri Poincaré, looking at how an idea is created and defining four phases: Preparation, brooding phase (the incubation), flash of inspiration (the illumination) and verification. All phases have chance in them, especially the first three.

We define three levels of intellectual activity: without chance, with moderate chance, and with a lot of chance. The limiting case of intelligent work—completely without chance, i.e. without uncertainty—is the *algorithm:* a set of work instructions. Working without chance or with little chance is *intelligence*. This is about solving tasks that involve uncertainty or/and are too large to be solved directly with work instructions, and involve uncertainty. The mechanisms for *creation*, the emergence of ideas, intermediate or normal stage, are associations or "bisociations". They are in the tradition of the thirteenth century wheels of Ramon Llull or the twentieth century method of Fritz Zwicky. Even the most extreme case on the edge of normal, the genius, is thus grasped and freed from the suspicion of the supernatural.

This clears the way for the creativity of the computer, after all we are also a kind of computer. The key is again chance, often as a stream of random numbers. With chance, music can be composed, pictures can be painted, poems can be written, 3D objects can be built. The computer can learn without "understanding"—as we do—and it can decide. We explain schematically how "decision-making" actually works—in a group of people in a conference room, programmed according to this model in the digital computer and in us humans ourselves. In this way we show that there is no free will—chance does not mean freedom—but that we nevertheless have alternatives and are not fixed. This only works with the various forms of chance. Chance brings extreme situations that are witty but nonsensical back to reality. This applies to "the donkey of Buridanus", who will certainly not starve to death, or the "Dome of Norton", which, however, turns out to be actual nonsense anyway.

The way the human mind has been classically imagined (or is still imagined) is the homunculus fallacy. We are a noisy computer, and noise as the source of randomness runs throughout the world and is as much a part of our creativity as it is of our "free" will. Our inner noise cannot be seen, hence the comparison:

Our brain is like a lake with gentle waves in light winds, sometimes with boats making ripples, sometimes with storms and surprisingly high waves.

References

Austin, James. 2003. *Chase, chance and creativity*. Cambridge, MA: MIT Press.

Davies, Gareth, 2017. Newtonian physics IS deterministic, sorry Norton. https://blog.gruffdavies. com/tag/the-dome/. Zugegriffen im Juni 2020.

Dobelli, Rolf. 2011. *Die Kunst des klaren Denkens*. München: Hanser.

Earman, John. 1986. *A primer on determinism*. Dordrecht: Springer.

Fauconnier, Gilles, and Mark Turner. 2002. *The way we think. Conceptual blending*. New York: Basic Books.

Hadjeres, Gaetan, et al. 2017. DeepBach: A steerable model for Bach chorales generation. arxiv.org/abs/1612.01010. Zugegriffen im Juni 2020.

Hartung, Gerald, ed. 2010. *Eduard Zeller, Philosophie und Wissenschaftsgeschichte im 19. Jahrhundert*. Berlin/New York: de Gruyter.

Hehl, Walter. 2016. *Wechselwirkung – wie Prinzipien der Software die Philosophie verändern*. Heidelberg: Springer.

———. 2019. *Gott kontrovers*. Zürich: Vdf.

Kant, Immanuel. 2019. *Kritik der praktischen Vernunft. Kritik der reinen Vernunft & Kritik der Urteilskraft*. Dachau: OK Publishing.

Koestler, Arthur. 1964. *The art of creation*. London: Hutchinson.

Kurzweil, Ray. 2012. *How to create a mind*. New York: Viking.

Lehmann, Alfred, and F. Bendixen. 1899. *Die körperlichen Äusserungen psychischer Zustände*. Leipzig: Reisland.

Mainzer, Klaus. 2007. *Der kreative Zufall*. München: Beck.

Mecke, Jochen. 2015. *Du musst dran glauben*. Diegesis. Uni Wuppertal.de. DIEGESIS 4, H. 1.

Meier, Carl. 1992. *Wolfgang Pauli und C.G. Jung. Ein Briefwechsel*. Berlin: Springer.

Mordvintsev, Alexander, et al. 2015. Inception. *Going deeper into neural networks*. AI googleblog.com.

Nicolic, Danko. 2018. Why do brains have spontaneous activity? Sapienlabs.org/why-do-brains-have spontaneous activity.

Poincaré, Henri. 1908. *Science et Méthode*. Paris: Flammarion. jubilotheque.upmc.fr.

Ruelle, David. 1994. *Zufall und Chaos*. Berlin/Heidelberg: Springer.

Segal, Michael. 2019. *Why the brain is so noisy*. Nautil.Us/issue/68.

Smale, Nick. 2005. Victor Vasarely. Artspace, Issue 23.

van Andel, Pek. 1994. Anatomy of the unsought finding: Serendipity. *The British Journal for the Philosophy of Science* 45 (2): 631–648.

Wagner, Astrid. 2005. *Kreativität und Freiheit. Kants Konzept der ästhetischen Einbildungskraft*. Acamia.edu/2186012.

Watters, Brett. 2019. How many bytes memory size is a humans DNA. Answer in Quora.com.

Weinberg, Bernhard. 1961. *A history of literary criticism in the Italian renaissance*. Chicago: Hathitrust.com und University of Chicago Press.

Zyga, Lisa. 2010. *Freewill is an illusion, biologist says*. Phys.org/news/2010-03.

Chance as the Foundation of the World

From Thomas Aquinas is an astonishing (from today's point of view) statement of chance in the thirteenth century, just as astonishing as chance was to Epicurus in the fourth century BC with the introduction of the clinamen, the spinning motion in everything. The "real" quivering motion is then discovered in 1827 by the botanist Robert Brown at the microscope and the "real" coincidence emerges in 1964 when the Northern Irish physicist John Stewart Bell publishes the inequalities named after him in quantum physics. Experimental tests then proved it with the most successful scientific edifice in human history, quantum theory: there is real chance. This was a short history of physical chance!

> **"It would be contrary to the perfection of things if there were no accidental events."**
> **Thomas Aquinas, Italian theologian and philosopher, 1225–1274, in Summa contra gentiles, Book 3, Chap. 74, c. 1260.**

This is an astonishing discovery of coincidence in the thirteenth century from today's point of view, just as astonishing as Epicurus' coincidence in the fourth century BC with the introduction of the clinamen, the spinning motion in everything. The "real" quivering motion is then discovered in 1827 by the botanist Robert Brown at the microscope and the "real" coincidence emerges in 1964 when the Northern Irish physicist John Stewart Bell publishes the inequalities named after him in quantum physics. Experimental tests then proved it with the most successful scientific edifice in human history, quantum theory: there is real chance. This was a short history of physical chance!

Thomas Aquinas gives freedom to man, because as a Christian theologian he juggles between the omnipotence of God and the free will of man. But Thomas Aquinas even gives a certain freedom to all of nature. Chance, though "contingent," is somehow also divine; God does not intervene in chance, but he might (Scarani 2015).

Our perceived problem with chance is the secular equivalent of Thomas Aquinas' theology: in this tradition of thought we juggle understanding chance as an illusion and

© Springer Fachmedien Wiesbaden GmbH, part of Springer Nature 2021
W. Hehl, *Chance in Physics, Computer Science and Philosophy*, Die blaue Stunde
der Informatik, https://doi.org/10.1007/978-3-658-35112-0_8

absolute chance without cause in principle. It has been placed in us by evolution to have to ask ever deeper questions about causes. We have shown that not being able to disentangle causes also seems like absolute chance. This does not imply a violation of causality; it continues to apply. It is simply forbidden to ask the dice why it created the six.

8.1   Noise as a Random Continuum and Motor

"Where do the forgotten thoughts go?"
Attributed to Sigmund Freud, Austrian psychologist.

We've already discussed different noise and different sources of noise as a kind of disturbance, if you look or listen closely enough. But noise is more. It is also the place where forgotten thoughts go, and the place where many a new thought arises. Whereas in forgetting, expressed in computer language, first the addresses of the places of the thoughts "noise", later the content. This may be the explanation that a conversation or hypnosis can still retrieve forgotten contents.

The usual, known and accessible to us humans noise is in a middle position on the scale of noise. At the lowest, finest level of randomness are ubiquitous quantum fluctuations and thermal fluctuations. Quantum fluctuations are tiny statistical fluctuations in the energy of a point in space for tiny times; the product of energy and time fluctuation is, according to the famous Heisenberg uncertainty principle, a certain multiple of Planck's constant h (which is also tiny). Thermal fluctuations are small random fluctuations of particles like atoms or electrons in thermal equilibrium. Each possible mode of particle motion receives the same tiny amount of energy to randomly tremble, rotate, or oscillate, the higher the temperature, the more. There even remains some of this energy at absolute zero that cannot be erased! In that sense, this book could have been called: *Everything rushes* in analogy to the already mentioned saying of Heraclitus *"Everything flows - panta rhei"*. We don't even have to go down into the depths of physics like that.

A wonderful visual example of noise in the sense of "small disturbances" is the sea or a lake with a light wind, now not seen as an explanation of interacting forces, but in terms of natural philosophy or simply as an experience.

The surface of the lake is a "lake" of coincidences: in the snapshot in Fig. 8.1 only spatially and in the temporal course spatially and temporally. After only one second, most of the wave crests would be somewhere else, or the individual pixels of the photo would already be different. If we now throw a stone into the lake, this random event generates waves which run outwards, have smaller and smaller amplitudes and finally get lost in the noise of the lake and finally in its thermal motion. In other words, the throwing of the stone into the lake introduces additional entropy into the lake, which becomes incorporated and inextricably mixed into the entropy of the lake. The event "throwing the stone" disappears into the noise underground. Disappearance means that there is a limit after which a single

Fig. 8.1 The undulating surface of a lake as a snapshot of the noise in a dynamic system with randomness. Evening atmosphere on Lake Zurich; the city is visible in the background. Image: Edith Geissmann

event can no longer be identified as an event in the noise, even with computer analysis (a chain of known events can be identified for longer).

The individual event with all its correlations, i.e. its identity, has irretrievably disappeared in the set of underground events (Fig. 8.2).

In classical mechanics everything is smooth and steady, even more than steady. Nature tries to make the trajectory of an object as little curved as possible (Hertzian principle). With noise it is just the other way round: The (microscopic) trajectory is no longer steady, no longer differentiable and in the amount of particles also no longer calculable.

The great form of noise, at least on earth, is then the waves on the oceans and the fluctuations of the currents in the ocean. But also random movements in the earth's crust, which as great coincidences cause volcanic eruptions, earthquakes and tsunamis, are random. And of course the noise in the cosmos goes on up to the largest structures, the galaxy clusters.

For us humans, the neurological noise in our brain is the most important thing - as a sign of life and as source material for creativity. Noise is the driving force for change in many forms, not only in evolution.

Fig. 8.2 The reverse
coincidence. The disappearance
of an event in the noise
underground and in the
unpredictability

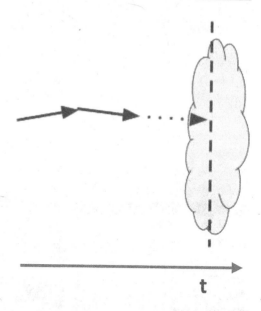

Noise determines the limits of predictability in nature as well. Almost an industry is numerical weather prediction, which since the availability of computers has also become an applied science: How far in advance can the weather be predicted in the chaotic world of atmospheric randomness? Today, at mid-range latitudes, it's about nine days. The absolute limit, according to American meteorologist Fuqinq Zhang, is about two weeks (Carroll 2019). That there is a limit to computing at all can be understood by our gradation of "very large":

humanly large – physically very large – mathematically very, very large.

Supercomputers typically calculate 10^{20} operations with given initial conditions in one hour, i.e. physically very large sets of numbers. However, the complexity of nature increases rapidly to completely different, inaccessible orders of magnitude. The combinatorial interaction of the many particles makes it "mathematically very, very large" and beyond computability. This applies to the world's weather, but also to the wave patterns of the small lake of Zurich, and even to the movements of water in a glass of water. Figure 8.3 illustrates the phenomenon: the initial values with their uncertainties in the cloud of noise are a major reason why predictions further into the future become increasingly uncertain and ultimately worthless. The only "computer" that can "calculate" this is the quantum computer of nature itself.

Noise is inconceivable underground, from which the new comes and in which much disappears. The American philosopher Charles Peirce speaks of the *"womb of indeterminacy"* in his 1887 paper *"A Guess in the Riddle"* (arisbe 2006).

This brings to mind the most famous womb of indeterminacy in the history of science, Darwin's already mentioned *"warm little pond."*

Fig. 8.3 The predictability limit, for example for weather forecasting. The certainty of the forecast disappears at this limit

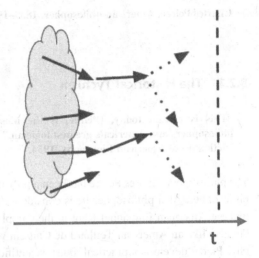

> "But if, oh what a big 'if,' if we could imagine a small warm pond containing all kinds of ammonium and phosphoric acid salts, plus light, heat, electricity, etc., so that .... etc."
> Charles Darwin in a letter to botanist Joseph Hooker, 1871.

For Darwin, this is even the possible place of origin of the first life, in today's terminology the beginning of (chemical) evolution.

We thus have three variants of images for the sources of coincidence:

- Philosophically, the *womb of indeterminacy* by Charles Pierce.
- Scientific or pre-scientific, the *pool* of Charles Darwin.
- Mathematically, the *oracle* of Alan Turing.

Surprisingly, noise exerts a fascination on people in many forms, especially as ocean noise, but also in its harshest form as piercing, computer-generated white noise, a stream of acoustic coincidences:

> "White noise is generally considered to be extremely psychoactive and trance-inducing [...] It can be used excellently as a background for hypnosis, self-hypnosis, meditation or for experimental work with trance."
> Hypnosis-Training-Seminar.com

8.2 Chance as a System: The Tychism

> "The endless variety in the world has not been created by law. It is not of the nature of uniformity to originate variation, nor of law to beget circumstance. When we gaze upon the multifariousness of nature we are looking straight into the face of a living spontaneity."

Charles Peirce, American philosopher, 1839–1914.

8.2.1   The Historical Tychism

"This is certain today, [Peirce] is the most original and versatile of America's philosophers and America's greatest logician."
Dictionary of American Biography. 1934.

The philosopher Charles Peirce has been a polymath, mathematician, logician, chemist and philosopher. As a philosopher he is considered the greatest in the USA. He is something like the American Immanuel Kant in the complexity of his philosophy and his language, but also like an American Teilhard de Chardin with the focus on love. In general philosophy, Pierce develops a practical, quasi-scientific approach that he calls pragmatism. One maxim of his doctrine is that (only) the practical and verifiable properties of an object matter. Pragmatism will become a mainstream of philosophy in the twentieth century.

Applied to chance and the fundamental (Fig. 2.2a) to chance, this means: Inaccessible causes, whether they exist or not, are unimportant. There is only the right side of the sketch, the source of chance. Chance is at the centre of his doctrine of tychism:

▶ Definition "Tychism is the teaching of the American philosopher Charles Peirce that absolute chance plays an active role in the universe."
"Tychism" in definitions.net

Peirce is dissatisfied with science at the end of the nineteenth century, but from a very different point of view than the sober Emil du Bois. Du Bois had tried to combine "mind" and mechanistic science. He had despaired of how "a thought or a sensation should arise from the movement of globules" - that *we shall never understand, ignorabamus!*" To solve the problem had lacked the entirely different science called computer science, and of course much else.

For Peirce, mechanistic science was too narrow and too unimaginative. He was fascinated by the efficacy of evolution and the incredible variety of plants and animals it had produced - through myriads of coincidences. Peirce's opening quote continues, in the spirit of, say, Benoît Mandelbrot and our earlier chapters, with the call for "looking closely."

"A day of roaming the countryside should actually bring that home to us."

Peirce is also a scientist. He knows Newtonian mechanics and thermodynamics at the time of the development of the concept of entropy. Above all, he is impressed by Darwin's doctrine of evolution as the dynamic emergence of new things by chance. It is amazing how he describes the origin of the universe (Reynolds 1996):

"In the beginning - infinitely far away - there was a chaos of unpersonal feelings, which being without connection or regularity would properly be without existence. . . . Thus, the tendency to habit would be started; and from this, with the other principles of evolution, all the regularities of the universe would be evolved." (Charles Peirce, CP 6.33, c. 1898)

It sounds like a poetic description of chaos in Fig. 4.1 and of the beginning of evolution. After the development of general relativity and the discovery of the expansion of the universe, the Belgian theologian and physicist Georges Lemaître came up with the idea in 1931 that at the beginning of the universe there was a kind of "cosmic urea". According to today's view, it was an ultra-dense, ultra-hot chaotic state (see Chap. 4).

Peirce sees many arguments for chance being a necessary, cosmological force:

1. The deterministic laws (it is still mainly about mechanics) can not explain the diversity of nature, see above quote. It is a more modern version of Epicurus' idea that atoms need disorder to move realistically (clinamen).
2. These deterministic laws cannot explain the tendency to grow, to increase complexity and to evolve, and this irreversibly in *one* direction. Many laws of nature can also be reversed and the reverse processes would also be possible, but they do not exist on a large scale. Thus, the theorem of the growth of entropy applies.
3. Evolution needs an indeterminate core as a starting point. It is the "womb of indeterminacy of the world".

To illustrate the diversity of the world we have included two figures, one "cosmic" and one terrestrial. The cosmic picture of Fig. 8.4 shows the astronomical structures in our extragalactic environment on the very large scale: Only faint structures can be seen, the subtleties of which then look literally random. Coincidence on a cosmic scale has played a part here. More colorful and familiar is Fig. 8.5, which shows the animal part of terrestrial life. The species and the individuals are shaped by chance. In Fig. 8.4 it is freewheeling chance; in Fig. 8.5 it is systematically accumulated chance with extensive blueprints. According to Peirce, chance is actively involved in the cosmos from the very beginning.

The scientific basis of Peirce's thinking is based, on the one hand, on understanding the principle of biological evolution, and on the other hand, on mechanics and its most mysterious concept, entropy, which fascinates Peirce. He cannot go further in science around 1900, only guess. So he misses quantum theory, which emerges a generation later and brings a new random component to science. He would have welcomed it. The discovery of the indeterminate quantum world has a profound effect on philosophy.

Comparable in meaning is the replacement of the then classical dualism "matter and mind" by the modern dualism "matter and information", i.e. physics and IT, but only a century later. Peirce tends to see matter only as a powerless form of mind. But he sees the development of the whole world as a great evolution, with and by absolute, i.e. inscrutable, chance. The generally accepted evolution of life is his great example of thought, but he extends the idea of evolution (wrongly) to the inanimate world. In the inanimate world, evolution is just a vague name for continuing development, but in biology it is a defined

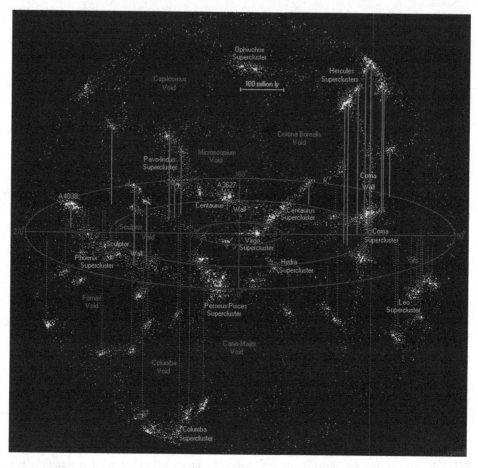

Fig. 8.4 Astrodiversity. Objects in the "near" vicinity of our Milky Way. The voids and superclusters up to 500 million light years away. Image: Nearsc, Wikimedia Commons, Richard Powell

process of construction. In our or Popper's terminology, we are dealing on the one hand with the development of inanimate world 1, and on the other hand with world 2' or world 2, both under the influence of chance. We can clarify this below.

The second basic mechanism of the development of the world after absolute chance is for Peirce to develop "habits". What he means by this becomes clear from the following quotation:

> **"The tendency to develop or generalize habits arises from one's own action, by the habit of taking habits itself growing".**

Fig. 8.5 Biodiversity (animal kingdom). Image: Animal Diversity, Wikimedia Commons, compiled from Wikimedia Commons images. Free Art License

The modern word for this is "self-organization". The term is also found in Immanuel Kant's *Critique of Judgment* of 1790, but only for the world of the living, and especially in the Swabian philosopher Friedrich Wilhelm Schelling, 1775–1854 (Heussler-Kessler 1994). In the modern sense we define:

▶ **Definition Self-organization is the emergence of a structure by chance when it occurs instantaneously, without a code or stored program, i.e., primarily in physics, but also in sociology.**

A physical example is the hexagonal structures of snow crystals, in social terms it is, for example, a blogger's group of followers on the Internet. It is ad hoc randomness. Evolution and the development of life itself is constructively accumulated randomness and has a very different dimension, both in the degree of complexity developed and in the framework of time, namely 4 billion years in place of spontaneity! But of course both are essentially random processes. Pierce understands the term *habit-taking*, adaptation through life, to include evolution itself, but also social evolution in society. "*Habit-taking*" means inventing, testing, perhaps rejecting. His cosmology is brief:

> **"Three elements are active in the world: first, chance; second, [natural] laws; and third, the adoption of habits."**

Charles Peirce calls his doctrine of the world with chance *tychism* after the Greek goddess Tyche, corresponding to the Roman Fortuna. Tyche is responsible in mythology for the changing course of history. She elevates and degrades people. Greek and Roman cities worshipped her as a kind of city patron saint. Figure 8.6 shows four Tyches from the fourth century BC.

Fig. 8.6 Four tyches as Roman city saints: Rome, Constantinople, Alexandria and Antiochus. From the treasure of the Esquilino Hill. British Museum, London. Image: Tychai Esquiline Treasure, Wikimedia Commons, Recruos/Jononmac46

The term tyche then became depersonalized and secularized into "chance"; according to Wikipedia, it even became a curse in case of misfortune, something like "damn" or even New German "fuck" (though the latter is not in Wikipedia).

Peirce describes the history of the world from the point of view of the Tychist very briefly thus:

> **"The evolution of the world is hyperbolic. .... The state before infinite time is chaos, tohu bohu, nothingness without any structure. The state in the infinite future is death, nothingness, in which law and order triumph completely and there is no longer any spontaneity. In between, we are [today] in a state with some spontaneity against the laws and more and more conformity to the laws."**
> **1891, in CP 8.317.**

At the time of Peirce's life in the nineteenth century, science had elevated determinism to a principle - sensibly so at the time. After all, there were still many laws to be discovered. Tychism was certainly a curiosity in that scientific world.

Charles Peirce was also an exotic as a person. His achievements as a philosopher are recognized only half a century:

> **"Charles Peirce was one of the greatest philosophers of all time."**
> **Karl Popper, Austrian-British philosopher, 1972.**

A particularly remarkable achievement of Peirce as a logician was his prediction in 1886 (!) that it would be possible to perform logical operations with electrical circuits. It would take until 1940 for Konrad Zuse to build an electromechanical computer with the Z2.

Pierce developed his doctrine of tychism and chance as the foundation of cosmology around 1891. The famous quote "God does not play dice", which has already been quoted several times, was written by Einstein in 1926 in letters to the physicists Niels Bohr and Max Born. In 1927, the physicist Werner Heisenberg formulated the uncertainty principle:

There are pairs of physical quantities, e.g. the location and the velocity of a particle, where only one quantity can be measured precisely, but then the other, the second quantity, necessarily becomes indeterminate.

This means that absolute chance was thus clearly introduced into science. Albert Einstein probably never accepted it.

8.2.2   Neo-Tychism: Absolute Chance in the Modern World

> **"The world is thereafter statistical and stochastic and not deterministic. Chance is itself the principle."**
> **Klaus Mainzer, German philosopher, born 1947, in "Creative Chance".**

If we want to understand a scientist of a past epoch, we have to "discount" the knowledge, take it back to his or her epoch. In the case of a philosopher, there is also the fact that he or she usually builds up a conceptual world of his or her own, which must be understood, and which is not free of contradictions in itself. It is not like in the natural sciences, where the outcome of an experiment inevitably corrects false assumptions (even in the case of an authority).

Charles Peirce is no exception. However, he is also a pragmatic scientist. He goes too far in his doctrine of tychism when he also sees the laws of nature as created by chance. Unless, of course, we see the adjustment of the natural constants that make up our universe as the result of a sophisticated random process, or think postmodernly that our universe is one of many universes. Our universe then just "happens" to have the set of natural laws that we observe.

In any case, there is a core of natural laws that are closely connected with each other with incredible precision, often to 12 valid digits. This core is certain as a whole. Astrophysics thus teaches us that the constants of nature, such as the speed of light, have had the same value over many billions of light years of distance and billions of years of time.

The principal innovation in "Neo-Tychism" is the dualism of matter and physics on the one hand, and computer science and mind on the other. Both worlds follow different laws.

We add to Pierce's arguments above:

4. The world of physics (world 1) has in itself fixed and precise laws, but evolves from the initial conditions and with chance. The order as a whole decreases, the entropy of the whole system increases unceasingly.
5. The world of computer science (world 2′) is characterized by further growing complexity, growing from "*habit*" and chance, as evolution shows it and as Peirce guessed it. These "gentle laws" of life are indeed built on with chance. Today, this world continues to grow primarily through the complexity that humans create in the form of "software." There is no end in sight to this growth.

Figure 8.4 shows the "random" distribution of astronomical objects on a very large scale, namely of galaxy clusters in our cosmic environment, i.e. world 1. Figure 8.5 illustrates the animal part of our biodiversity, i.e. a part of world 2′, created by chance and in a random composition. The difference in scale between the two areas is about $1:10^{25}$.

We have to add to point 3 "initial amount of chaos" that there are many chaos nuclei, for example the waves of a lake or the movement of sperms or the noise in the brain. The idea is compatible with quantum theory and quantum fluctuations "way down" in the smallest dimensions, with kinetic theory of gases, and with solid state physics. We add this as a sixth thesis:

6. There are infinitely many chaotic nuclei in the world, on earth as well as in space, which are closely connected ("entangled"). The largest random areas on earth are the atmosphere and the oceans.

   They are all small and big *"wombs"* for newness, for chance, for spontaneity.

As the seventh thesis, we describe the world's mode of operation:

7. The course of the world is determined by chance within the given framework of the laws of physics. At least in world $2'$, the natural part with biology, chance is dominant.

Figure 4.9 of the mountain stream just impressively describes this partly chaotic flow with the boundary conditions, the flowing of the water in the rock bed. Thereby it is unclear, whether in the coincidence and in connection with the laws of physics there is a kind of definite inclination, which finally necessarily results human life. We have already discussed the question in the chapter Evolution, e.g. for the eye:

Assuming the same or very similar initial conditions, would evolution converge to the similar lens eye or similar life forms?

And the very big convergence question:

Assuming a repeat of the Big Bang, would the cosmos evolve similarly? Would life evolve similarly?

This leads to the inherently almost absurd question:

Does chance alone, or together with the laws of physics, have intrinsic properties, a propensity or "propensity" after all?

Maybe randomness is just abstract and just allows nature to reach ergodicity, i.e. all corners of the universe's possibility space.

Together with Peirce's classic three points above - the incredible diversity in nature, the fixed direction of the arrow of time in an apparently reversible world, and the necessity of chaos as the starting point of everything - we have altogether formulated a coherent cosmology with operational randomness.

We try and invent for illustration an analogue of the formation and course of the cosmos. Let it be a fictitious country that gets a railway network. It is like Europe in the first half of the nineteenth century, USA around 1830 and China around 1870. There is no railway in this country yet. Now the railway is introduced (Fig. 8.7):

(a) The technology of the railway is predetermined and fixed. The track width of the rails is defined. If electric, voltage and current frequency are defined.

This corresponds to the laws of nature in cosmogenesis.

(b) But it is free (and largely random) where the tracks are laid.

This corresponds to the development of the material world 1.

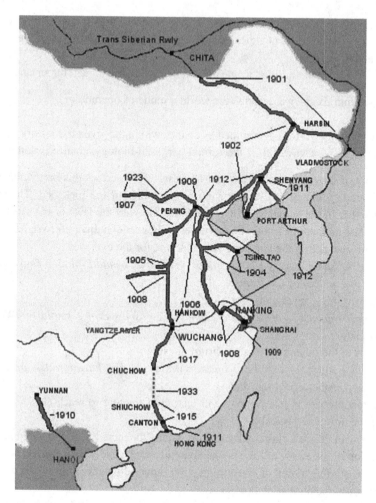

Fig. 8.7 The early railway network of China. To compare cosmogenesis with the construction of a railway network. Image: KCRC Early Network of China, Wikimedia Commons, Mosr

(c) Now more and more trains are used and timetables are created. Widespread tourism is created. This illustrates the intangible part of the railway. *It corresponds to the world 2'.*

Neo-Tychism is defined in outline. As a caveat, the author is not aware of any corresponding modern citation that advocates neo-Tychism in this form. But the importance of chance is acknowledged by modern philosophers, notably by the German philosopher of science Klaus Mainzer (b. 1947) in his mathematics-influenced, highly recommended book, "Der kreative Zufall" (Mainzer 2007).

The often-mentioned US philosopher Daniel Dennett, arguably one of the world's leading philosophers, popularly puts it this way (Dennett 2003), in a rather lengthy quote:

"Isn't it true that whatever isn't determined by our genes must be determined by our environment? What else is there? There's Nature and there's Nurture. Is there also some X, some further contributor to what we are? There's Chance. Luck. This extra ingredient is important but doesn't have to come from the quantum bowels of our atoms or from some distant star. It is all around us in the causeless coin-flipping of our noisy world, automatically filling in the gaps of specification left unfixed by our genes, and unfixed by salient causes in our environment."

This results as a "world formula":

We are "Nature + Nurture + X", with chance X.

According to the Merriam Webster dictionary, *nurture* here means:

"the sum of environmental factors that express the behavior and characteristics of an organism".

Now, "nature", like "nurture" itself, has also come into being with and through chance. If we use the quasi-mathematical expression $A(X)$ of the function A in the sense of "property A depends on X", this results in a complete world formula:

We are made up of nature (by chance X) + nurture (with chance X) + chance X.

Or in short: **We are We (X),** we are a work of chance, in this moment and historically. We ourselves, our history, our mountains, our forests, all animals and plants, our weather, our seas.

Chance is more than a stopgap: the laws of nature are the framework for the flow of chance.

8.2.3   Summary of the Chapter

For classical physics, chance is disruptive in the individual and is tamed in the crowd with the help of mathematical statistics. But chance is more: it is actively part of the world from the beginning. Chance in miniature is everywhere: it is noise. A beautiful example of visible fluctuations in space and time is a lake (or the sea) with its waves. If a stone is "accidentally" thrown into the lake, the triggered waves eventually disappear unidentifiably into the noise. The disappearance of an event in the noise is the inverse coincidence. Conversely, new things come out of the noise, such as straight ideas in the mind. The omnipresent random noise gives a principal limit to the predictability of events, for example in weather forecasting. The noise as a stream of randomness is always behind the determined processes. Sources of noise are everywhere and thus also the possibility of spontaneous change.

Chance is an integral part of the cosmological foundation of the world. The American philosopher Charles Peirce was the first to see chance as the basic principle of the world from its beginning. The usual laws of nature provided too little diversity, he believed. It is the same reasoning that led to the introduction of the clinamen in ancient atomic theory.

According to Peirce, chance is already introduced into the world at the beginning in a chaotic beginning. Today we know the "Big Bang" for it. He sees the growth of complexity in the course of development as evolution, but also wrongly as a Darwinian evolution of inanimate nature. Peirce thinks that the laws of nature also arise through a kind of evolution. The dualism matter to information shows today that worlds 1 and 2′ evolve differently: World 2′ (life) builds blueprints, the core of the physical laws of nature is fixed with incredible precision. Thus, we have cosmology:

- *The chaotic core emerges with the "Big Bang" at the beginning.*
- *World 1 (the inanimate world) continues to evolve passively with chance, but based on the exact laws.*
- *World 2′ (life) begins to build up blueprints from this by chance, which can be passed on and thus grow in complexity. It is Darwinian evolution with chance as the driving force.*
- *Intelligent life is evolving that can draw up blueprints and build devices to execute them (the computers and their software).*

We call this doctrine neo-Tychism in honor of Charles Peirce. Of course, Charles Peirce did not yet have any idea of the genetic processes taking place. Without the fundamental absolute chance it does not work. Not for the cosmos, but also not for us humans as individuals, even if we almost cannot grasp it.

We literally bring the understanding of chance to a formula according to philosopher Daniel Dennett:

We are made up of nature (by chance X) + nurture (with chance X) + chance X.

Thereby X stands for coincidence. Or in short: **We are We (X)** in pseudo-mathematical notation: We are a product of chance.

References

Arisbe. 2006. *Charles Peirce – A guess at the riddle.* arisbe.sitehost.iu.edu.
Carroll, Matthew. 2019. *The predictability limit.* Sciencedaily.com. Zugegriffen am 15.04.2019.
Dennett, Daniel. 2003. *Freedom evolves.* New York: Viking Books.
Heussler-Kessler, Marie-Luise. 1994. *Schelling und die Selbstorganisation.* Berlin: Duncker und Humblot.
Mainzer, Klaus. 2007. *Der kreative Zufall.* München: Beck.
Reynolds, Andrew. 1996. Peirce's cosmology and thermodynamics. *Transactions of the Charles S. Peirce Society* 32 (3): 403–423.
Scarani, Valerio. 2015. The universe would not be perfect without randomness. Arxiv.org/abs/1501. 00769. Zugegriffen im Juni 2020.

Chance in Human Life

9

Chance is actually a foreign body in our thinking, at least the chance that one cannot question or see through. The above quotation has a noble and witty equivalent in Einstein's saying *"the old man does not play dice" (see above).*

We first compare the life positions with chance in the classical religious worldview and in the likewise classical scientific thinking of the Enlightenment.

Traditional Religion

In our traditional religious world-thinking, the cosmos is orderly, optimally put together, and "created." The causal chains we see end (or begin) in fabled pasts with powerful gods who are usually, but not always, on our side. The Abrahamic religions have abstracted many of the physical characteristics of the gods, but the human or inhuman traits have remained. To this end, there are procedures for securing the gods' goodwill - through total submission, through sacrifice, and through special intercessors such as the saints. When the possibilities are exhausted (or the measures are insufficient), then the only thing left to do is to accept fate, especially death, and to interpret misfortune as a punishment or a test. If something seems like a coincidence, there is still the possibility of seeing a higher being behind it. That is what these two well-known sayings express:

> **"Chance may be the pseudonym God takes when he doesn't want to sign"**
> **Théophile Gautier, French writer, 1855, sometimes attributed to Albert Schweitzer.**

---

[1] *Imho* is internet slang meaning "in my humble opinion".

© Springer Fachmedien Wiesbaden GmbH, part of Springer Nature 2021
W. Hehl, *Chance in Physics, Computer Science and Philosophy,* Die blaue Stunde der Informatik, https://doi.org/10.1007/978-3-658-35112-0_9

and

> "Le hasard, en définitive, c'est Dieu" Chance is in reality God.
> Anatole France, French writer, 1906.

With this way of thinking, chance seems to be tamed in our lives; it is only an incidental, slightly disturbing effect anyway. The central thing is the higher beings.

Enlightenment

The Enlightenment has replaced quasi-human gods at the beginning of causal chains with rational causal chains that are recognized or suspected until they are lost in the distant past. Reason is now the great arbiter. The causal chains of events that affect us can now be traced further and more reliably, but at some point the dark and unknown does begin. At the same time, order is being created in the foreseeable realm by the natural sciences, and coincidence hitherto suspected there is being removed, for example with the successful prediction of the course of the sun, moon and planets. Friedrich Nietzsche asks:

> "What is evil? Three things: chance, the uncertain, the sudden."
> Nachgelassene Fragmente, Religion, 1888.

In the same fragment, Nietzsche describes the history of culture as a decrease in the fear of chance: "Cultur, that means learning to calculate, learning to think causally, learning to prevent, learning to believe in necessity." Chance does not fit, and as absolute chance it does not exist in determinism. But it exists, of course, in practical life, in nature, scientifically tamed as statistics, half-tamed and somewhat uncanny in the form of entropy, and untamed in the action of evolution, which creates something new and greater with chance.

The economic taming of chance is an industry in its own right, the insurance industry as the "chance industry". Its mission is defined by *"eliminating the risk of one individual through the contributions of many with similar risk"* (according to Wikipedia article Insurance (collective), pulled July 26, 2020).

The beginnings date back to before the Enlightenment in the solidarity community of the guilds. At the end of the eighteenth century, the first insurers emerge on a mathematical basis. Since the risks for coincidences have become greater and greater since then, from the nineteenth century onwards another kind of "taming" emerged, now for the insurers themselves: the reinsurers. More or less well tamed is chance in another chance industry, the casinos - at least from the casino's point of view, as long as the bank wins.

The effect of chance through a fire, through the loss of a ship or a gambling stake is drastic, painful and real. Chance as an abstract mechanism of an evolution had a hard time against the official, almost tangible creation story. The resistance has been formidable. Chance does not fit properly with religion and not at all with a reason that wants to explain everything down to the smallest detail. Uncertainty does not fit with the apparent certainty of people at the end of the nineteenth century. Indeterminacy relation and entanglement in quantum physics in the twentieth century are in total contradiction with the reason of a Laplace. Chance in life is obvious, but only as a disturbing side effect (personal death is the most disturbing thing). The resulting dismissive attitude to chance is not so different from the religious point of view.

Neo-Tychism

But there is "real" chance, absolute and fundamental, also for our life. Chance is almost everywhere, the smooth and simple is rather the exception. The laws of nature are valid and all processes are causal, but much, very much is random. Chance provides the diversity in nature and in our lives, beginning with our birth, our "being" in philosophical parlance. It's just as Nietzsche said, "Instead of authority by a god or reason, *chance, the uncertain, the sudden* rule." We define:

▶ **Definition A process is tychistic if it works only with absolute chance.**

Two "tychistic" events limit our life: birth out of the sea of coincidences with our parents and our genetic combination, and death, where we sink back into the noise. No one planned the events (family planning is not enough). Of course, one can hope for a plan in the background, but this would be unscientific and only a human analogue. Our example above, the throwing of a stone into a lake or river and the fading of the waves in the thermal noise and random eddies of the water is an analogous, simple example to the teleology of life: from the scientific-technical worlds 1 and 2' there is no meaning to our life. Only the direction of developments is given.

Also biological evolution, the scientific prime example of tychism, gives no sense in itself. The blueprints of organisms have arisen by accumulated chance, but are now bastions against chance precisely as complex but sufficiently stable programs for life. For the continuation of evolution it is necessary that at least one of the accumulated successful blueprints is not lost and at least one spark of life continues to exist in a living being with this plan. This is the minimum requirement of life. To do this, not all species need to be kept alive, and to do this, not all individuals need to be kept alive as long as possible. Ethical demands such as "preservation of species", "protection of individual life" or even "protection of human dignity" have nothing to do with evolution. They are all world 3' constructs, outside of the technical. There is no such thing as natural law.

Of course, there are simple animal species, such as the coelacanth or the lamprey, that have existed in the world for hundreds of millions of years: Why didn't evolution settle for that? In the evolution of commercial software, it is the pressure of new customer requirements that forces the code to evolve. Evolution seems to have a *posteriori a* direction and a tendency, a "felt" propensity, and the direction is us humans, us as social beings (humanly speaking) or "us" as creators of the next generation of IT, digital IT (technologically speaking).

For us individuals and our life this does not help. We are thrown into life and from now on it "happens", according to the biological rules, the social rules of fellow humans and with continuously intervening chance.

A German confirmation song by journalist and songwriter Jürgen Werth (born 1951) says wonderfully soothingly:

> *Never forget: you are no accident, no freak of nature.*
> *You are a brilliant thought of God, you are precious!*
> . *That you live was not your own idea. And that you breathe, no decision of yours*

The text sounds good and feels wonderful, but only in the security of human terms: We are many coincidences, from the beginning. We are whims of nature, if that image is allowed. The word "whim" means 'caprice, fancy, sudden turn of the inclination of the mind' after the Online Etymology Dictionary. The equivalent German word "Laune" comes from *luna*, from the moon, and its transformations. We do not control a large part of the world in principle. In principle, there is no sign of security, neither for the species "man" nor for us as individuals. It is true: it is not our idea that we are there.

Existentialism

This thrownness as an attitude towards life is not new, but already emerges in the nineteenth century with the Danish philosopher Søren Kierkegaard, 1813–1855.

> **"...I learned this from Socrates. I want to make people aware that they are not wasting their lives and ruining them."**
> **Søren Kierkegaard, Diaries II.**

Kierkegaard is thus a precursor of Jean-Paul Sartre, whose maxim *we* have already mentioned *"to bring man into possession of himself"*. It is the doctrine of existentialism with its main representatives Jean-Paul Sartre, Albert Camus and Simone de Beauvoir in France around the middle of the twentieth century. It is an attempt for a philosophy of moral and meaningful life with chance. Sartre writes "l'essentiel, c'est la contingence":

> **"The essential is the accidental. I want to say that if you want to define existence, it is not the necessary. To exist is simply to be there".**
> **In: La nausée (the disgust), 1938.**

The motto of existentialism is: *l'existence précède l'essence,* translated roughly: Existence precedes essence, or by way of explanation: our essence only emerges in the course of life in the encounter with the accidents of life and is not predetermined at birth.

Coping with life in a world of relentless, even absurd, coincidences is no longer the focus of many modern philosophers' work today. Existentialism is not a main direction of philosophy. Rather, so-called analytical philosophy is pursued with concrete individual topics, such as feminism or the ethics of modern medicine. But every individual has to cope with his life and "his" or "her" coincidences. Existential questions are therefore at the heart of many literary works. The website *goodreads.com* maintains an impressive list of 100 works of world literature on existential themes:

Abe, Beckett, Bradbury, Camus, Dostoevsky, Hesse, Huxley, Kafka, Kundera, Nietzsche, Rilke, Saint-Exupéry, Salinger, Sartre, Shakespeare (Hamlet), Voltaire, Wolf, and many more.

As a side note, there are also "tychistic" acts to mention. For this, Jean-Paul Sartre defines the *actes gratuits*:

▶ **Definition** **An acte gratuit is an arbitrary act with no comprehensible motivation, usually violent.**

Such an absolutely random act is a symbolic rebellion against what is perceived as oppressive determinism and the compulsion of causality or, religiously, a rebellion against the will of God. The classic example is a murder in André Gide's 1914 novel *The Vatican Cellars*: the "hero" throws an unknown old man off a moving train without motive.

The real deal with chance in life is then sociology, psychology, religion: is it luck or bad luck, punishment or reward, sign or coincidence? When it is simply chance and you understand it, you become more humble to the course of history.

A form of humility in the face of chance is expressed by comforting phrases such as: *"C'est la vie!"*, *"Dumm gelaufen"* or internationally *"Shit happens"*.

According to Wikipedia, the latter is "part of the vernacular" and is not considered vulgar. Those who want to say it genteelly can also exclaim *"Stercus accidit!"*, the Latin version. It's shorthand for "tychism," so to speak.

The folk wisdom *"Every man is the architect of his own fortune"* is a superficial view of life according to our discussion of "free" will. But there is a wise folk addition:

> **"Everyone is the architect of his own fortune, but chance is always blowing the bellows".**
> **From Fliegende Blätter, humorous German weekly magazine, 1944.**

There are chance situations in life that are beyond the frivolities of colloquial speech: The death of a neighbor or the proximity of one's own death, serious illness, danger to life. The philosopher Karl Jaspers introduced the concept of *Grenzsituationen* or borderline situations in 1919 (Jaspers 1919). These are situations in which man inevitably and unmistakably comes up against the limits of his life. Much of this is a direct coincidence and it is important to think:

> **"I must die, I must suffer, I must struggle, I am subject to chance, I inescapably entangle myself in guilt".**
> **Karl Jaspers, German-Swiss philosopher, 1883–1969.**

The Norwegian painter Edvard Munch (1863–1944) painted several versions of the matching picture that made him world famous and stands at the beginning of the style of Expressionism in painting, Fig. 9.1, the "Scream". It corresponds exactly to the spirit of

Fig. 9.1 The image of man in a borderline situation due to a threatening event. "The Scream" by Edvard Munch, painted in 1910. (Image: Edvard Munch The Scream, Wikimedia Commons, Google Art Project)

threatening chance and conveys the insight into the life of the soul at the moment of despair. It is the impression of a life situation in which we can do nothing but surrender. Our life "*happens*," it happens.

References

Jaspers, Karl. 1919. *Psychologie der Weltanschauungen*. Berlin: Springer.

Conclusions

10

> "Chance, for Lichtenberg, is nothing more than a form of connection in the world, connected with objective lawfulness."
> Dorothea Götz, in *Georg Christoph Lichtenberg* (Goetz 1984).

The small, hunchbacked but ingenious eighteenth century physicist Lichtenberg probably suspected it: Chance is not just a superficial disturbance in the otherwise regulated course of nature, but it is built into the foundation of the world. Even the ancient atomists, the inventors of the atomic idea, knew it: without chance there is no living world. From the beginning of the world, chance has been responsible for the diversity in the cosmos, in the astronomical world of stars and galaxies as well as on earth.

The discovery of biological evolution by Darwin was the first clear indication of this: it is chance that can bring the new into the world, indeed only chance. But actually chance as absolute coincidence without comprehensible justification is an unpopular and elusive thought. This was one reason for the opposition and antagonism Darwin experienced, even to the point of sneeringly asking whether he was descended from the ape "on his father's side or on his mother's side." It is unfortunate or even foolish to speak of the "theory" of evolution in terms of something questionable. Evolution is nature's method of creating complexity and driving the direction of the world's development. The laws of physics cannot do this, at most entropy can—but that is also part of chance. It is a quantity that measures chance in terms of order or disorder. However, evolution does not produce "meaning of life", neither for a species, nor for an individual. Evolution only "wants" to go on.

This makes chance a problem for us humans: it is uncontrollable and thus a threat that we have to deal with. One possibility for this is to see in chance the effect of a higher power. Chance is interpreted religiously—as a mercy or a punishment. The Enlightenment has given us the false security of being able to understand everything in principle. Real existing chance deprives us of this confidence. Our lives are chance, guided by the framework of

© Springer Fachmedien Wiesbaden GmbH, part of Springer Nature 2021
W. Hehl, *Chance in Physics, Computer Science and Philosophy*, Die blaue Stunde der Informatik, https://doi.org/10.1007/978-3-658-35112-0_10

natural laws. The hardest consequence of this insecurity in principle was drawn by the philosophers of existentialism like Sartre and Camus, not so popular today, but the basic ideas already: *shit happens* and we have to live with it (for this expression, see text).

We find that chance need not be so absolute: For the effect to be a coincidence, it is enough that we can no longer ask about the cause: Why is it so? For instance, to the dice *"Why didn't you roll a six?"* But of course every result in dice rolling always has a cause, but this cause is by definition inscrutable behind a wall. We show physically and mathematically how this "not being able to ask" comes about: There are smooth transitions between visible and transparent on the one hand and inscrutable chaos on the other. We can see this in the throwing of a stone into a lake (how the triggered waves fade away) or in the predictions of weather (when the predictions become blurred after one or two weeks). Being physically unpredictable and mathematically unpredictable go hand in hand. What is no longer predictable or calculable is, by definition, noise. Noise is almost everywhere—in water, in the air, in our brains, at the moment of sexual reproduction. Expanding the concept with the mathematician Mandelbrot, the grasses of a meadow, the clouds in the sky, and the waves on the lake are noisy—that is, there is randomness everywhere, often in broken geometries as "fractals." Of the total information describing, for example, a fully grown tree, only a tiny part is hereditary information.

We encounter the first noise in its most concentrated form at the Big Bang, at the creation of the universe. But today there are such "pools of chance" everywhere, in which chance arises before our eyes and, on the other hand, events pass away. Visible are the waves of the lakes and the sea, invisible is the fluid flow in the fallopian tube during fertilization or the neuronal noise in the brain. In death, we too become noise altogether. Our structural information disappears and our physical matter spreads out over the world.

We have seen the interplay of chance and necessity in the physical example of the mountain stream: The streambed gives the framework, the starting point and the end point of the flow, but the path of the individual water molecules is indeterminate. A framework for the world as a whole is given by conservation laws such as energy, momentum, angular momentum and charge, which are deeply built into the internal symmetries of the world. These symmetries like a displacement in time or in three-dimensional space or a rotation in space also determine the behaviour and properties of elementary particles.

Another important restriction is the "Pauli principle", named after the Austrian physicist Wolfgang Pauli (1900–1958). It determines which atoms are possible, but also which larger structures, and applies to most of the real world. Thus, the processes of the world are limited and the physical possibilities are given, in which the course of the world with its coincidences and causal chains take place.

The elementary analogy of the introduction of a railway system in a country demonstrates the laws (such as the rail standard and the number of locomotives as a "law of conservation") and the evolving random world that takes place on the rails.

The result is our world with a lot of conserved and with living chance: the animal species and the mountains, for example, have conserved chance, our life is active chance. We are, according to Nobel Prize winner Manfred Eigen, active participants in the great world game,[1] but the game goes deeper than Eigen probably thought: We are not independent homunculus players in the random world, but the game of chance goes on in our minds as well.

In addition, an observation about the great processes of chance. The future is unpredictable precisely because of the coincidences, but if we look back, the observation is repeated that everything is exactly as it had to be for us to exist: the sun, the earth, evolution, our immediate ancestors. It is thus given meaning to everything *a posteriori* (it is the so-called anthropic principle). Thus, chance can appear to have a propensity to lead to a certain goal—and it is not clear how strong the propensity, of the universe and the foundations of physics is to lead to life through and with chance—and whether such a "propensity" exists at all. Emotionally, all stages of the history of mankind look like a lawful and necessary sequence to today: Stone Age, Bronze Age, Iron Age, industrialization with fossil energy, now with solar energy, with digitalization, but also with climate change. There is no guarantee for the future. The good chain of chance could continue according to law—or come to an end with our generation. It is conceivable that there is "zero propensity", i.e. no inclination at all, and that chance only has the task of ensuring access to all conceivable areas of the possibility space.

Chance grounds the new in our world, especially our creativity, visible in serendipity, but also otherwise more or less in human ideas and decisions. We analyze the (so-called) free will. It is an illusion that chance gives free will. Chance only generates alternatives, but has no responsibility. It is also an illusion that we have the brain (matter) here, the ego (mind) there, which makes use of the brain and commands it. The I is identical with the brain and its "software". It is wonderful that we feel free will, but actually we think with it in an infinite regression (homunculus effect). The idea here is to define as the center of the ego a homunculus in the brain that commands, and in it is again a homunculus that commands the first homunculus, and so on. There are wonderful theoretical borderline cases of decision such as the millennial *ass of Buridanus* and the modern (but fallacious) *dome of Norton*. Chance solves both puzzles realistically because "everything rushes."

The chapter on the cosmological significance of chance is based on the thoughts of the American philosopher Charles Peirce around 1898. He was an outsider at the time, but today he is considered by the philosopher Bertrand Russell to be the most important philosopher in the history of the USA. Peirce understands chance as a necessity for the world, for diversity and development in general, and this before the discovery of quantum theory with uncertainty principle and entanglement and before chaos theory. With these discoveries, Peirce would find himself vindicated. In homage to him, we use his term for the doctrine of the fundamental importance of chance, the term tychism from the ancient Greek word for chance and luck and the corresponding Greek goddess Tyche Τύχη.

The aim of the book is to show the reader that we not only live in a world of chance, but that chance intervenes in the world more deeply than we think. To this end, we use a world

---

[1] Manfred Eigen in the preface to *Das Spiel. Naturgesetze steuern den Zufall*. Piper, 1975.

formula that goes back to the American philosopher Daniel Dennett and the English play on words "nature and nurture":

We are "Nature + Nurture + X", with chance X.

But since both nature and education have come into being with and through chance X, we add to the formula.

We are nature (by chance X)+nurture (with chance X)+chance X.

Briefly in pseudo-mathematical notation: **We are We (X)**—we are chance.

The doctrine of chance *per se* should be a part of every curriculum for physicists, biologists, philosophers, theologians, for everyone.

The next time the reader sees the waves of a lake or the sea, may he or she remember chance as the foundation of the world.

References

Goetz, Dorothea. 1984. *Georg Friedrich Lichtenberg, Biographien Band 49*. Leipzig: Teubner.

Glossary

Abiogenesis emergence of life from dead chemistry.

Acte gratuit a completely senseless act.

Agent Detection the human ability to see living beings even in the inanimate.

Agile Software Development a software development almost without planning that reacts quickly to changes (coincidences).

Algorithm a concrete calculation rule.

Anthropic principle almost trivial fact that all conditions for the existence of the observer must be present.

Association (chance) mechanism to create new by connecting.

Consciousness accessible totality of functions that control an organism. In humans, a corresponding app associated with speech.

Big Bang Core of the cosmos in the formation of the universe with highly concentrated coincidence.

Bisociation (random) mechanism to create something new by connecting two areas.

Buridan, ass of a philosophical, extreme decision situation.

Clinamen artificial quivering motion of atoms in ancient atomic theory.

Dome of Norton a (fallacious) philosophical extreme decision situation.

First-person shooter the world in total first person perspective (wrong English loanword).

Entropy measure of the degree of disorder in a system.

ergodic process that can fully use all degrees of freedom.

Evolution in the narrower sense, a process for generating something new by chance.

Existentialism philosophical direction that focuses on life with chance. Especially around 1950 in France.

Bottleneck effect change in gene distribution when the population is reduced.

Fractal very irregular curves, surfaces or bodies with effectively broken dimensions.

Free will fictitious sense of freedom of choice.

Fate fictitiously fateful coincidence for people.

Homunculus effect the impression that the ego and brain are separate. Homunculus is actually a small person.

Incubation core phase in idea generation with a lot of chance.

Intelligence ability to solve tasks under ambiguous conditions.

© Springer Fachmedien Wiesbaden GmbH, part of Springer Nature 2021
W. Hehl, *Chance in Physics, Computer Science and Philosophy*, Die blaue Stunde der Informatik, https://doi.org/10.1007/978-3-658-35112-0

Capillary waves small waves formed by surface tension. Random waves in low wind.

Coal dust a trick to introduce chance into a situation. The speck of dust destroys order.

Creativity the ability to create something new.

Laplacian demon a spirit that knows absolutely everything at this moment and thus also knows the future.

Maxwellian demon a mind that can see and react to individual atoms and molecules.

Metaheuristics a general solution procedure.

Monster wave an exceptionally high marine wave caused by coincidence.

Near field Area of chaotic behavior at the origin of a disturbance.

Naturalism a doctrine of the world without supernatural and inexplicable processes.

Neural networks adaptive and adaptive filters.

Oracle in Alan Turing's case, a black box that provides transcendent mathematics, e.g. true random numbers.

Panentheism view that God is both in the world in everything and outside the world.

Pantheism view that God is in everything in the world.

Propensity a chance with preferential direction, e.g. a loaded die.

Pseudo-random number Apparent random number for which, however, a generating algorithm exists.

Qualia (plural), Quale (singular) subjective experience of sensations.

Rastrigin function a constructed function for the simulation of many stable species.

Noise a continuous stream of random signals, temporal or spatial or both.

Regression to the mean the trend to normal after the occurrence of a special event.

Butterfly Effect an arbitrarily small random change in the initial conditions causes a large change in the process.

Self-organization formation of complex structures without a stored blueprint.

Serendipity an unexpected serendipitous find. The English word is common.

Soliton a wave that travels through a medium almost unweakened and in a constant form.

Statistical the term "statistical" refers to random data collected. From the French *statistique* (state) science and the Latin *statisticum* concerning the state.

Stochastic Dependent on or influenced by chance. From the ancient Greek. στοχαστ (stochastikos) or "presumptive".

Strategy system of rules to act in unclear situations.

Tesla Sphere a device for generating random effects with plasma discharges.

Turing mechanism a physical-chemical mechanism devised by Alan Turing for generating random patterns.

Tyche Greek goddess of fate, synonym for chance.

Tychism philosophical doctrine which states that chance is a fixed and necessary part of the cosmos.

Watchmaker effect the effect that we can tell by feeling whether an object is world 1 or world 2'.

Covert inhibition (latent inhibition). filtering mechanism in the brain for unimportant stimuli.

World 1 according to Karl Popper, the inanimate world (physics).

World 2 according to Karl Popper, the world of the subjective.

World 2' the updated world 2 as a world of all objects with construction plans.

World 3' the hypothetical world of the spiritual, such as art and love.

Literature

Chaitin, Gregory. 2007. *Algorithmic information theory – Some recollections.* arxiv.org/pdf/math/0701164.pdf. Zugegriffen im Juni 2020.

Eigen, Manfred, and Ruthild Winkler. 1975. *Das Spiel. Naturgesetze steuern den Zufall.* München/Zürich: Piper.

Goetz, Dorothea. 1984. *Georg Friedrich Lichtenberg, Biographien Band 49.* Leipzig: Teubner.

Jaspers, Karl. 1919. *Psychologie der Weltanschauungen.* Berlin: Springer.

Madhavji, Nazim. 2011. *In memory of Meir Lehman.* pleiad.cl/iwpse-evol/keynote/slides.pdf. Zugegriffen im Juni 2020.

Peitgen, Heinz-Otto, and Peter Richter. 1986. *The beauty of fractals: Images of complex dynamical systems.* Heidelberg/Berlin: Springer.

Index

© Springer Fachmedien Wiesbaden GmbH, part of Springer Nature 2021
W. Hehl, *Chance in Physics, Computer Science and Philosophy*, Die blaue Stunde
der Informatik, https://doi.org/10.1007/978-3-658-35112-0

Printed in the United States
by Baker & Taylor Publisher Services